D1611269

PHOTOGRAPHIC SENSITOMETRY

WILEY SERIES ON PHOTOGRAPHIC SCIENCE AND TECHNOLOGY AND THE GRAPHIC ARTS

WALTER CLARK, EDITOR

PHYSICAL AND PHOTOGRAPHIC PRINCIPLES OF MEDICAL RADIOGRAPHY
Herman E. Seemann

OPTICAL AND PHOTOGRAPHIC RECONNAISSANCE SYSTEMS
Niels Jensen

ELECTRO-OPTICAL PHOTOGRAPHY AT LOW ILLUMINATION LEVELS
Harold V. Soule

AERIAL DISCOVERY MANUAL
Carl H. Strandberg

PRINCIPLES OF COLOR REPRODUCTION
J. A. C. Yule

TELEVISION FILM ENGINEERING
Rodger J. Ross

LIGHT-SENSITIVE SYSTEMS
Chemistry and Application of Nonsilver Halide Photographic Processes
Jaromir Kosar

PHOTOMICROGRAPHY, Vols 1 and 2
Roger P. Loveland

THE PHOTOGRAPHIC ACTION OF IONIZING RADIATION
R. H. Herz

IN-WATER PHOTOGRAPHY
L. E. Mertens

SPSE HANDBOOK FOR PHOTOGRAPHIC SCIENTISTS AND ENGINEERS
Society of Photographic Scientists and Engineers
Edited by Woodlief Thomas, Jr.

PHOTOGRAPHIC SENSITOMETRY—A SELF-TEACHING TEXT
Hollis N. Todd

In Preparation:

PROPERTIES OF THE PHOTOGRAPHIC IMAGE
Joseph H. Altman

INFRARED PHOTOGRAPHY, *3rd Edition*
Walter Clark

PHOTOGRAPHY IN THE VISIBLE AND NEAR VISIBLE SPECTRUM
Allan G. Millikan

Photographic Sensitometry

A SELF-TEACHING TEXT

HOLLIS N. TODD, Professor Emeritus
The Rochester Institute of Technology

A Wiley-Interscience Publication

JOHN WILEY & SONS, New York • London • Sydney • Toronto

Library of Congress Cataloging in Publication Data:

Todd, Hollis N
Photographic sensitometry.

(Wiley series on photographic science and technology
and the graphic arts)
Includes bibliographical references and index.
1. Photographic sensitometry—Programmed instruction.
I. Title.
TR280.T6 770'.7'7 76-22657
ISBN 0-471-87649-6

Printed in the United States of America

10 9 8 7 6 5 4 3 2 1

Preface

Sensitometry is a branch of experimental engineering physics that comprises methods of finding out how photographic-sensitive materials respond to exposure and processing. It originated near the end of the 19th century in the work of Ferdinand Hurter and Vero Driffield, two English amateur photographers with scientific backgrounds. At that time, gelatin-based silver halide photographic products were just coming into the market in great variety. Hurter and Driffield recognized the need for suitable test methods, primarily to test the claims of different manufacturers that their materials were faster than those of their competitors. During the past century their methods have been refined and expanded, so that there now exist well-developed techniques that lead to an understanding of how photographic films and papers behave.

Sensitometric methods have their most obvious application in manufacturing photographic products, where it is necessary to test samples of product to see whether or not their speed and contrast, for instance, meet the requirements and to test new methods for improved results. Sensitometric methods are also basic to consumer tests of a variety of potentially useful materials for selection of one that best serves the consumer's needs, and are necessary in the quality control of photographic processing. The methods are clearly necessary when photographic methods of data acquisition are used, as in mapping, photoreconnaissance, astronomy, radiography, and so on.

In addition to these applications of sensitometry, a study of the subject is essential to any photographer who wishes to go beyond the casual snapshooting stage of photography and who therefore wants to substitute rational procedures for trial-and-error ones. Ansel Adams and Minor White have applied the methods of sensitometry to their Zone System, which John Dowdell and Richard Zakia have formalized in their Zone Systemizer.

The subject matter of this book is fundamental. It deals with the concepts that are basic to an understanding of the photographic process and with the main factors that contribute to the excellence of the image.

ABOUT THIS BOOK

This is not a usual textbook, nor is it in any sense a test, although it may resemble one at first glance. It is in fact a carefully constructed tool for learning, to be used at your convenience and own pace. The subject matter has been broken down into small elements, all of which when added together form a complete package of understanding and skill. The learning method is based on well-established learning theory and evidence, which show that learning is most effective when the process actively involves the student.

Each section begins at a level that is suitable for anyone who has had even a limited experience in photography. No special knowledge is assumed. Starting from this point, the necessary concepts are developed, almost as if you had a private tutor by your side. Testing of the material shows that everyone is able to work through the material easily and practically without error. At every step you can check your progress.

Most persons find this kind of learning interesting and enjoyable. It is especially important not merely to *read* the material but faithfully to follow the directions found near the beginning of each section. Practice with the material is built in, so even if some of the items seem too easy or repetitive, it is important not to skip any. It is a good idea not to work with the material over too long a stretch; half an hour at a time is probably enough.

We begin with four sections dealing with the basic concept of variation in measurement. This material is fundamental to all data from experiments, and to the application of measured values to practice.

Sensitometry proper begins with Chapter 2, which introduces the basic measurements in the photographic process. The student not yet familiar with logarithms should, before working with Chapter 2, complete Appendix A, Sections A1–4.

Chapter 3 briefly develops the skills needed to work with the logarithmic scales found everywhere in photography. Chapters 4–9 treat in detail the sensitometry of negative materials, including such concepts as negative contrast as affected by exposure and development, gamma and contrast index, latitude, and film speeds. Chapters 10 and 11 deal with similar concepts for photographic papers. In Chapters 12 and 13 application of the preceding material is made to the process of making photographic prints. Chapter 14 completes basic sensitometry with a treatment of the relationship between the original subject and the final print image.

In the appendixes are found instructional materials on some of the equipment commonly used for exposing and measuring photosensitive materials. Appendix Sections C and D clarify the confusing subject of light-measurement concepts and methods.

HOLLIS N. TODD

Rochester, New York
March 1976

Acknowledgments

The assistance of the following persons is deeply appreciated: Dr. Todd Bullard, Provost of the Rochester Institute of Technology, for the grant that supported the preparation of this work, Professor John Trauger for assuming the burden of handling many of the problems involved in the preparation of the text, my colleagues John Compton, Ronald Francis, Thomas Hill, Leslie Stroebel, and Richard Zakia for their support and advice, and several thousand students of many ages and backgrounds for their patience and good humor, as well as their often blunt criticism.

H.N.T.

Contents

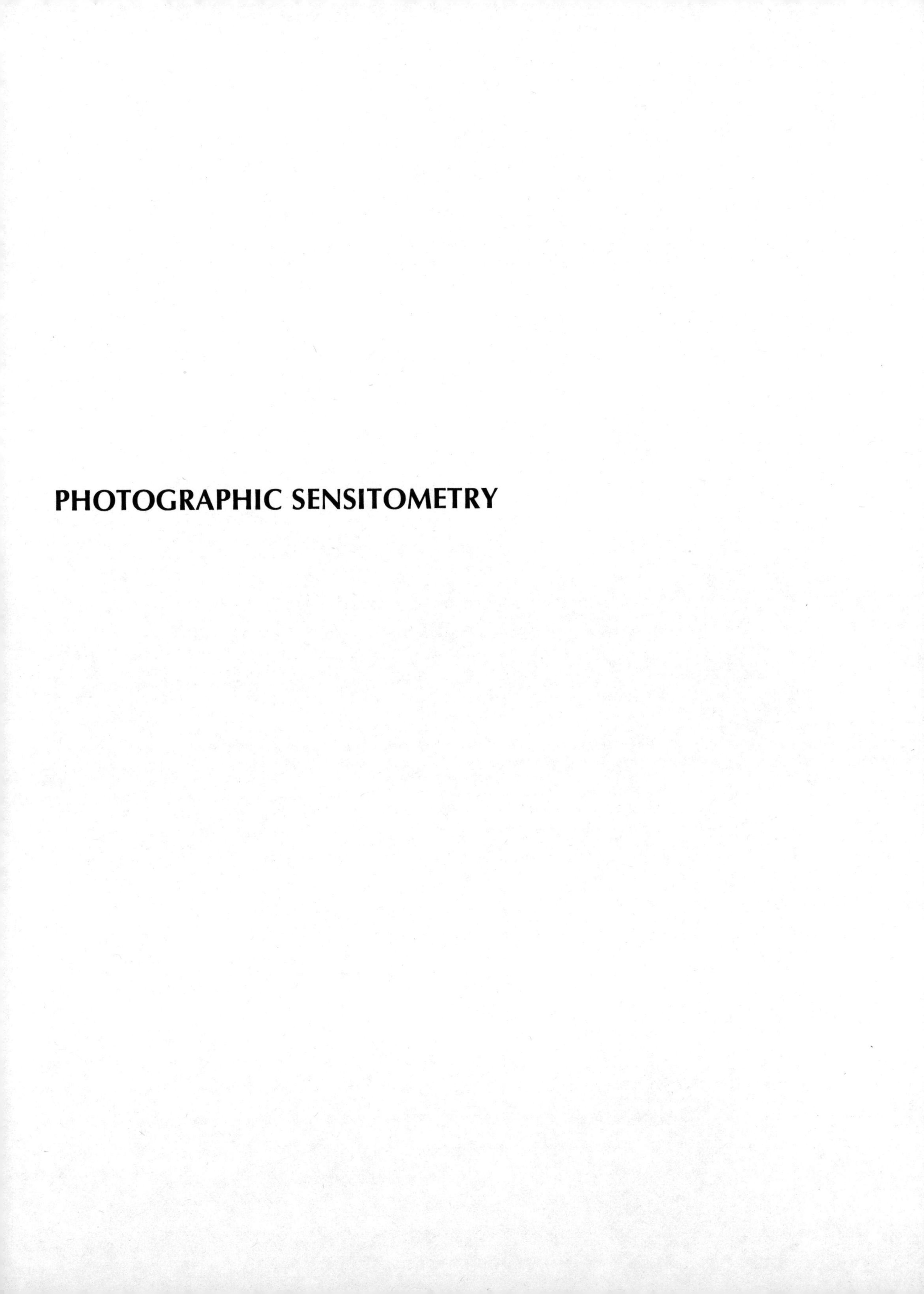

PHOTOGRAPHIC SENSITOMETRY

CHAPTER 1
Variability
SECTION A
INTRODUCTION

All photographers are, at least to some extent, involved with numbers that relate to the many factors in the photographic process. Some of these numbers are counted, for example, 100 sheets of paper in a package and 12 exposures left on a 20-exposure roll.

Most of the important numbers, however, are *measured.* Examples are a film speed of ASA 64, a lens focal length of 50 mm, a development time of 5 minutes, a liter of developer, and a shutter time of 1/100 second.

Counted numbers are potentially exact. If you were to find only 19 frames on a 20-exposure roll, you would be somewhat astonished, and perhaps you would protest. If you found, on the other hand, that a processing bath temperature was 25.5 C when it was supposed to be 25 C, you would hardly be amazed.

Measured numbers are *never* exact. In every statement that includes a measured value, there is always some uncertainty. When you pour out a liter of developer, the volume is *never exactly* a liter. Sometimes the uncertainty is suggested, as in the direction to process color film at "25 C ± ½." More often the uncertainty is left for you to guess at, and to estimate yourself, as in the direction to develop film for "8 minutes."

The following programmed material deals with some of the basic principles that will help you to deal with the variation in the measured values common to all photography. The need is almost obvious. If for example, a box of film is marked ASA 125, you must be able to cope with the hard truth that the film speed is not *exactly* 125 and with the even harder truth that the exact speed (if indeed it exists) can never be determined. Similar statements would apply to light meters, shutters, and so on.

Instead of striving for an impossible exactness, it is more fruitful to measure the variation that in fact always occurs. The basic methods used in such a study of variation are so important that we devote to them the first four sections of this book.

WHAT YOU WILL LEARN

If you work carefully with the material in this section, you will be able to:

1. Tell from a set of measurements how carefully the data were obtained;
2. Identify from a large number of similar data the pattern of variation;
3. Distinguish between sets of data that are controlled by chance and other sets of data.

DIRECTIONS: Cover about 2 inches of the right-hand margin of each of the following pages in turn with a sheet of paper. Read carefully each of the numbered statements or paragraphs. *Write* your response to the blank left in each statement on the cover sheet. Move the cover sheet down to expose the correct answer in the margin. Continue until you have finished the section. Work at a rate that is comfortable for you, and no longer than you find pleasant. Begin when you wish.

1. A very large (200-liter) tank is used for developing motion-picture film. A precise thermometer gives a reading of 20.2 C at the top of the tank. Elsewhere in the tank you would expect the same thermometer to give ____________ reading.
   the same, a different

   a different

2. We *know* that the temperature is not exactly the same throughout the tank. Thus we expect differences, or *variation,* in the temperature readings if they are carefully done. If a thermometer with only 2-degree scale intervals were used, we might record the "same" value for different trials. But such a result would only mean that the measuring process is too crude to reveal the real temperature __________ .

   differences, or variation

3. A precise shutter tester is used to measure the actual operating time of 10 shutters, all set at 1 second. A reasonable prediction is that the instrument will show that the 10 shutters are all __________ .

   different

4. If all of the values were the same, we would suspect that the tester could not show the differences that were really present.

   To take a different example, the scales of many photoelectric light meters resemble the sketch that follows. If five persons used five such meters to take a reading from a subject and all reported the same value, we would have reason to suspect that the measurements were made __________ carefully.
   very, not very

   not very

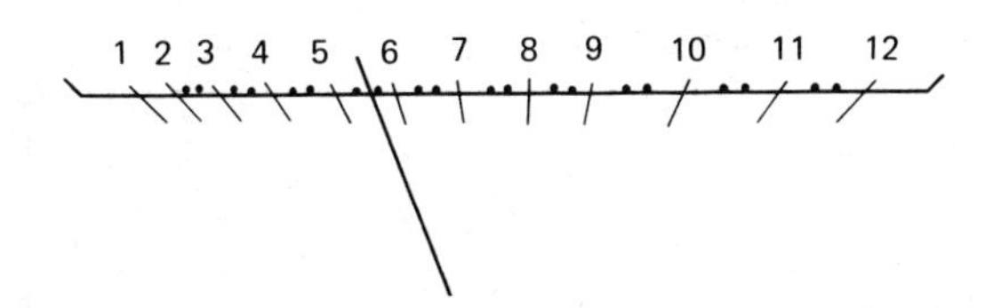

5. If, on the other hand, a similar experiment gave these data: 10.4, 10.7, 11.0, 11.2, and 10.9, we would have reason to believe that the measurements were made ___________ carefully.
very, not very

very

6. In an air-conditioned film processing plant, a technician made humidity measurements frequently for a day. He reported always the same value—50%. A reasonable judgment is that the humidity in fact was ___________ .
constant, changing

changing

We know that in the preceding example it is impossible for the humidity to stay the same over the whole day. Our best judgment is that the humidity changes but that the measurement method is not good enough to reveal the differences that truly exist.

The preceding examples illustrate that we do not need to experiment to find out whether differences exist in measured numbers. We *know* that differences exist. The important questions to be answered from experiments are these:

1. What *kind* of variation occurs?
2. How *much* variation occurs?
3. Can we tolerate this kind and amount of variation?

In what follows we deal with the first two of these questions. The third question cannot be answered in general terms because it involves matters such as the desired quality of the product or process and the demands of the customers, all of which are most complicated.

Patterns of Variation

7. If an honest (unweighted) coin is honestly tossed many times, we expect that the number of heads and the number of tails will be nearly ___________ .

equal

8. Such coin tossing is a process controlled by *chance*. By "chance" is meant that group of causes over which we have *no* control. In a chance-controlled process, the fall of the next toss ___________ be predicted.
can, cannot

cannot

9. In a chance-controlled process any single event can never be predicted. What *is* predictable is the *pattern*, in the long run. In an "infinite" (indefinitely large) series of tosses of a single coin, the percentage of heads will be nearly 50%. Such a percentage we call the *relative frequency*. "Frequency" means the count of the number of occurrences of a given kind. For *all* possible tosses of an honest coin, the relative frequency of heads is the same as that of tails. Let us identify a "tails" as 0 and a "heads" as 1. Now you may make a bar graph to show the simple pattern of the expected results for many tosses, using these axes:

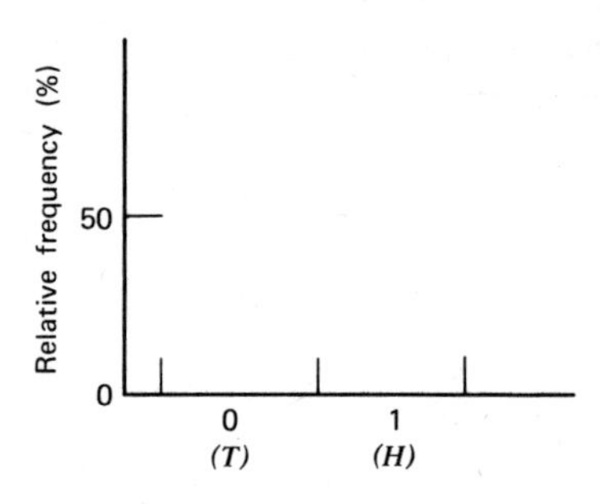

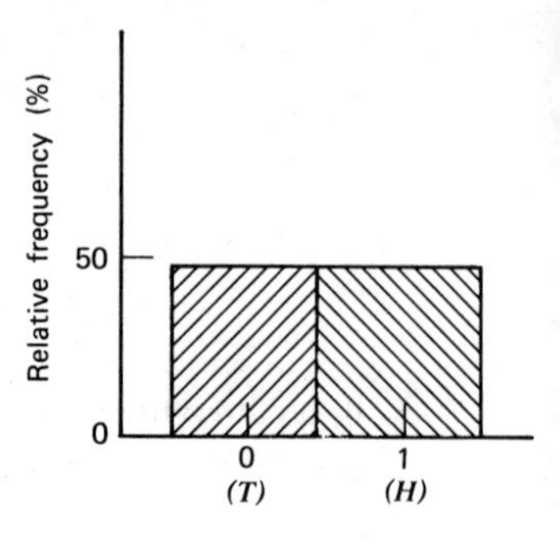

A correct graph is shown in the margin.

10. You have made what is called a "frequency *histogram.*" It shows the *pattern of variability* to be expected in the chance-controlled process of tossing a single coin.

 In any real experiment with a single coin, there is a limited number of trials. In such a *sample,* it would certainly be a rare outcome to find the number of heads *exactly* equal to the number of tails. We call the set of *all possible* tosses the *population.* In the population (an "infinite" number of trials) the number of heads for an honest process will equal the number of tails. In a sample with a limited number of tosses, say only 10, the actual result for a chance-controlled process might resemble this:

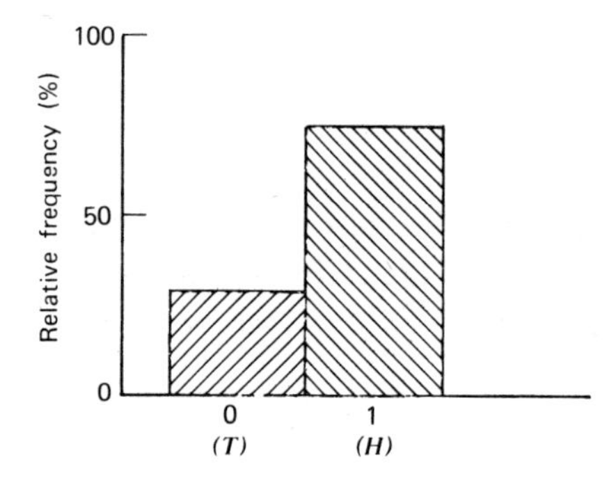

If, on the other hand, a *large* number of tosses, say 100, gave this plot, it would certainly be sensible to conclude that the tossing process __________ chance
was, was not
controlled.

was not

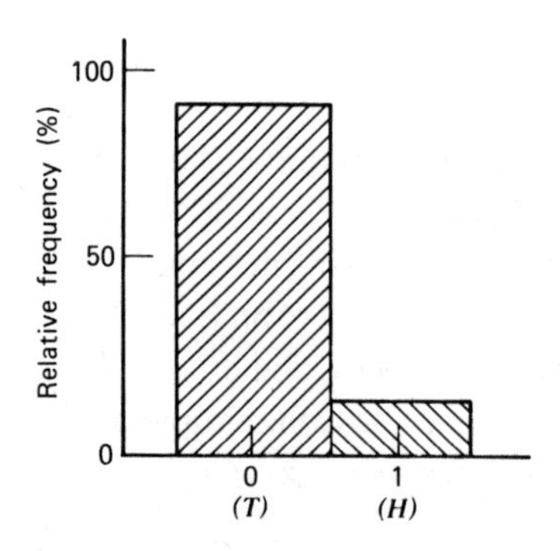

11. We symbolize the number of trials—the *sample size*—by the letter *n*. If *n* is small, fewer than perhaps 30, we can hardly expect that it will reveal the true pattern of the population—all possible trials. Only if *n* is large—say a few dozen—we can expect the pattern shown by the sample to come close to the pattern of the population.

    We run an experiment with many trials to find out whether the process is controlled only by chance. If the resulting pattern disagrees with that expected from chance causes, we then look for the cause of the unexpected result. If the pattern resembles the one to be expected from a chance-controlled process, we do not look for causes, since *chance* by definition implies causes that we ___, can ______ discover. cannot

    cannot

Let us expand the coin-tossing experiment a little: we toss two coins simultaneously, and score the number of heads as 0, 1, or 2 for each toss. The *long-run* expected results are:

Number of Heads	*Relative Frequency*
0	0.25 (25%)
1	0.50 (50%)
2	0.25 (25%)

12. Use the data above to make a frequency histogram (bar graph) of the expected results:

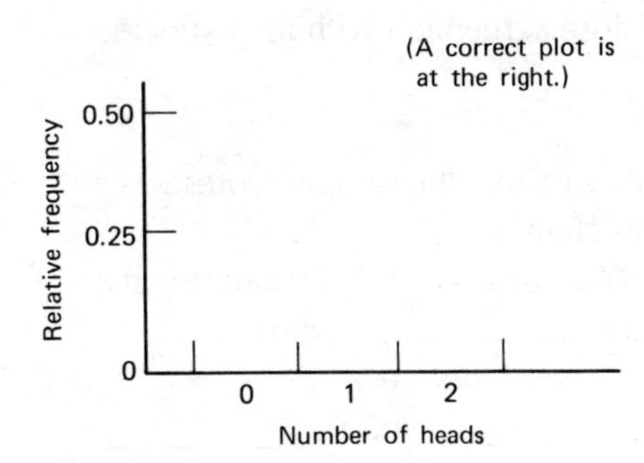

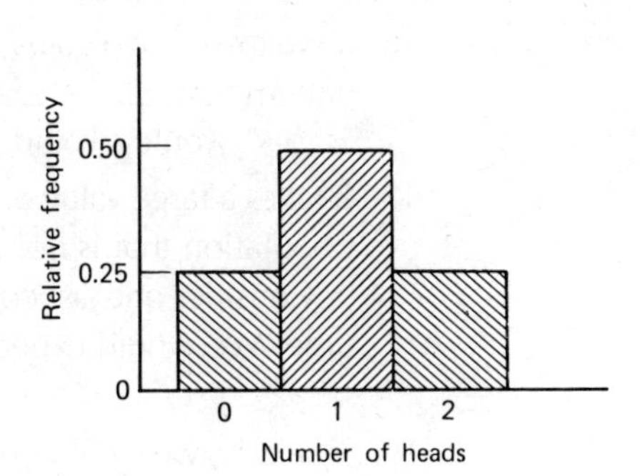

13. If we tossed two coins 60 times and obtained a plot like that shown below,

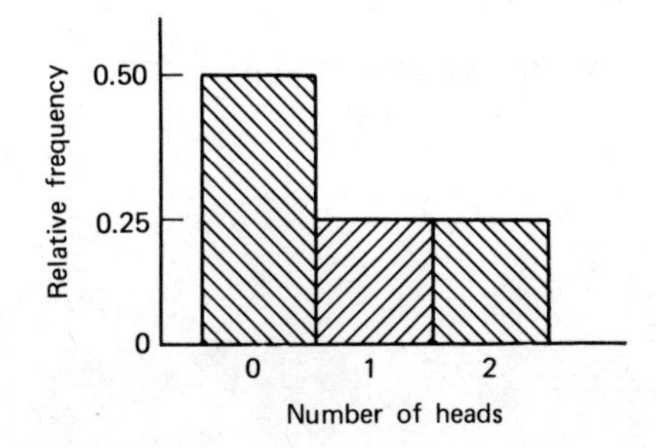

our conclusion would be that the tossing probably ________ chance controlled.
(was, was not) — was not

14. In item 13, n was a relatively large number. Thus when the pattern is clearly different from the one to be expected (item 12) we conclude that some cause other than ________ is causing the abnormally large number of tails. — chance
15. If another experiment involved tossing two coins 100 times, and we obtained this result: our conclusion would be that the process probably ________ chance controlled.
(was, was not) — was

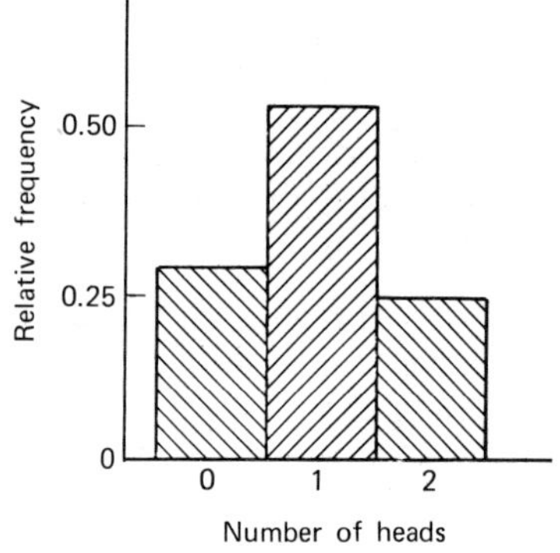

The pattern above is not *exactly* like the expected one in item 12. It is not, however, sufficiently different to compel us to believe that the process is not chance controlled.

16. If we tossed two coins only six times ($n = 6$), close agreement with the expected pattern ________ be likely.
(would, would not) — would not
17. It takes a large value of n for us to be reasonably sure that the sample represents a population that is *not* altogether controlled by chance.
Consider one last coin-tossing experiment. We repeatedly toss 10 pennies at a time. We would expect to get 10 heads per toss ________ often.
(very, not very) — not very
18. Similarly, we would expect to get *no* heads on a single toss of the 10 ________ often.
(very, not very) — not very
19. It would be far more likely that on a given toss we would get a few heads and a few tails. The most frequent result, in the *long run,* would be ________ number of heads and tails. — an equal, or the same
20. For the *population,* all possible trials, the expected pattern (for a chance-controlled process) is shown below.

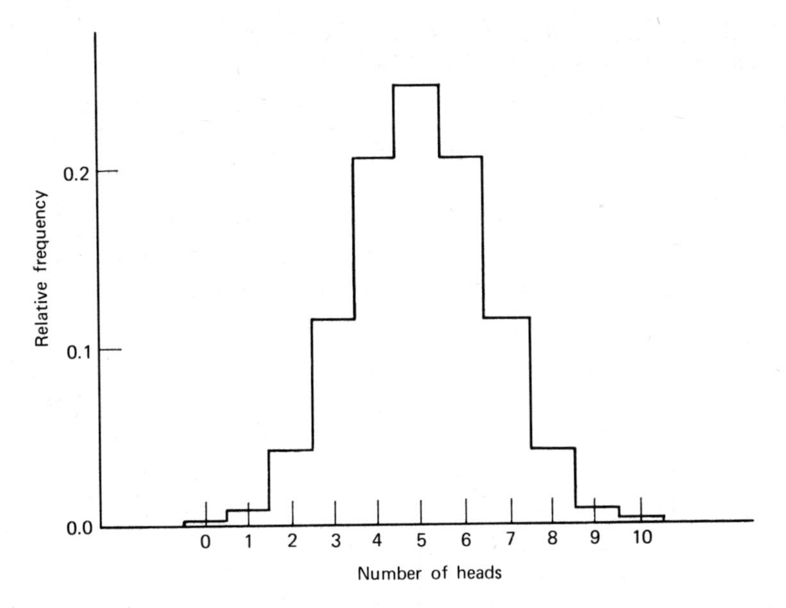

Suppose that in a sample of $n = 250$ tosses of 10 coins at a time you obtained the following result:

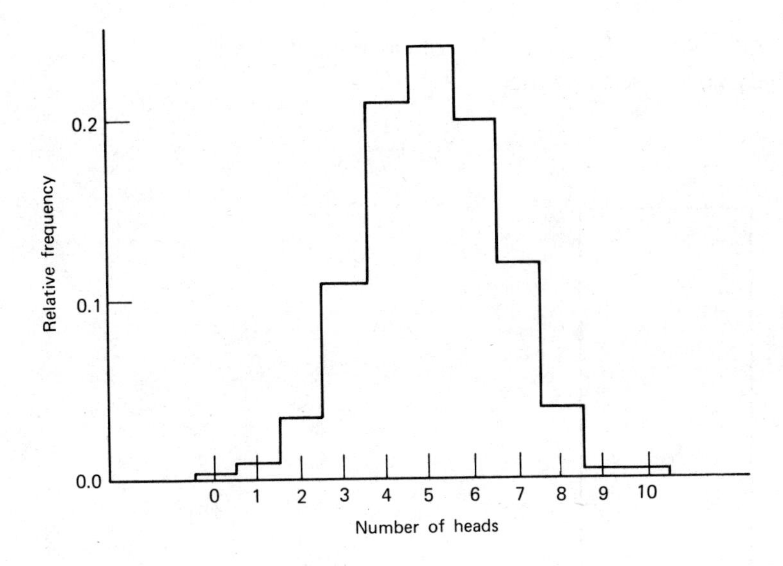

A comparison of the sample plot with the expected frequency histogram at the top of this page causes us to conclude that the process probably ________controlled by chance.
is, is not

is

Although the sample plot is slightly different from the expected one, the differences are too small to force us to conclude that the process is affected by some cause other than chance.

21. If, on the other hand, you obtained this result in a similar experiment with large n: you would be forced to conclude that something other than chance ______ affecting the results.
is, is not

is

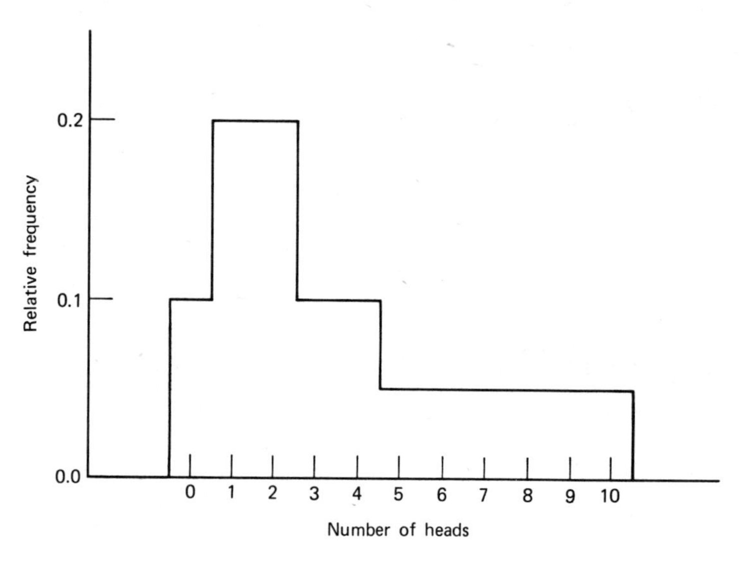

22. The following result from a large number of trials would convince you that the tossing process ______ entirely governed by chance.
is, is not

is not

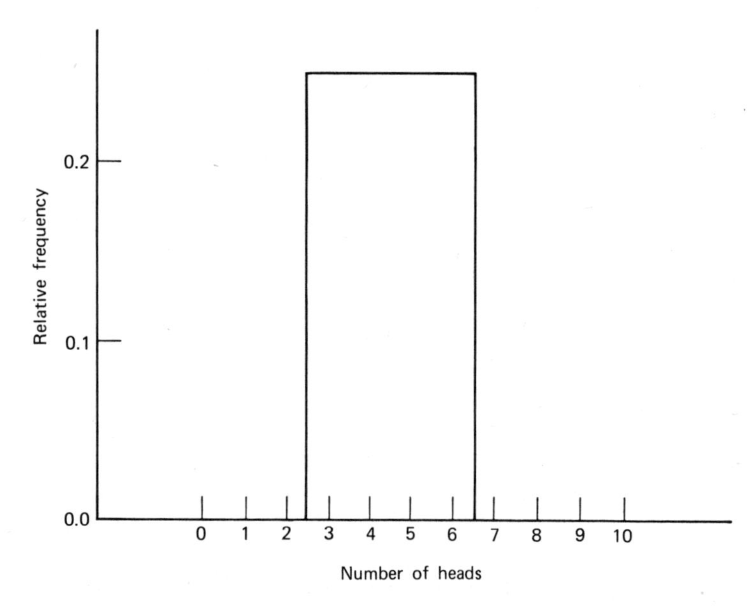

23. Note the change in the chance-caused pattern as the number of coins was increased from one to two and then to 10.

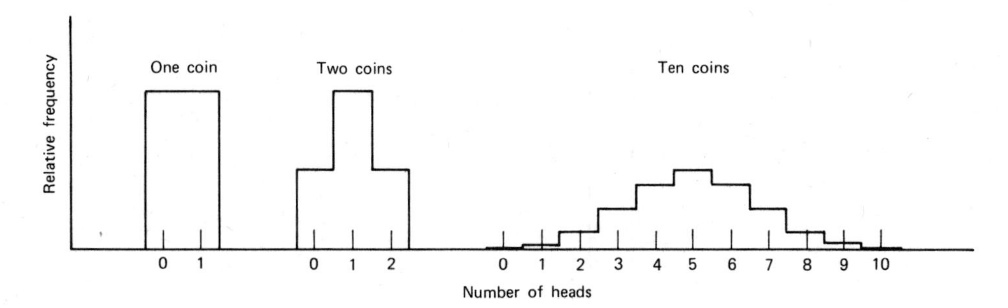

The change in the pattern shape is associated with the increase in the *number of different possible results,* which was only two for the single coin, __________ for two coins, and 11 for 10 coins. Now in a *measurement* process, the number of different possible results is practically unlimited. Thus the expected pattern for such a process is an extension of the 10-coin case to its limit—a smooth curve like the one that follows:

three

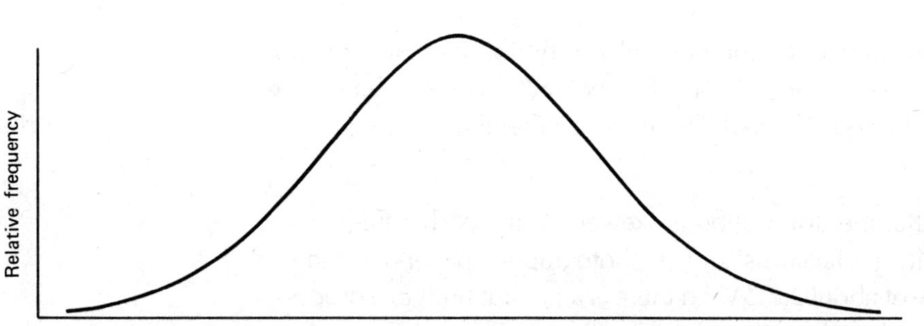

24. This graph shows the pattern to be expected from a large number of *measurements* when the results are controlled by __________ .

chance

25. The pattern in item 23 is commonly known from its shape as the "bell" curve. The technical name for this pattern is the "normal" distribution. It is a very important pattern because it is descriptive, not only of measurement processes, but also of production processes such as filling of a container to a desired weight or volume, cutting of film to desired size, the production of film with desired speed. All such processes, if chance controlled, will with large values of n give plots that resemble the __________ curve of distribution.

normal

26. For such processes, any considerable difference between a plot of real data and the normal distribution indicates that the process is not wholly controlled by __________ .

chance

27. Sometimes the reason for the difference between a frequency histogram of real data and the normal distribution is obvious. The following, for example, is a plot of the density (roughly, blackness) obtained when many samples of a glossy photographic paper were fully exposed and developed:

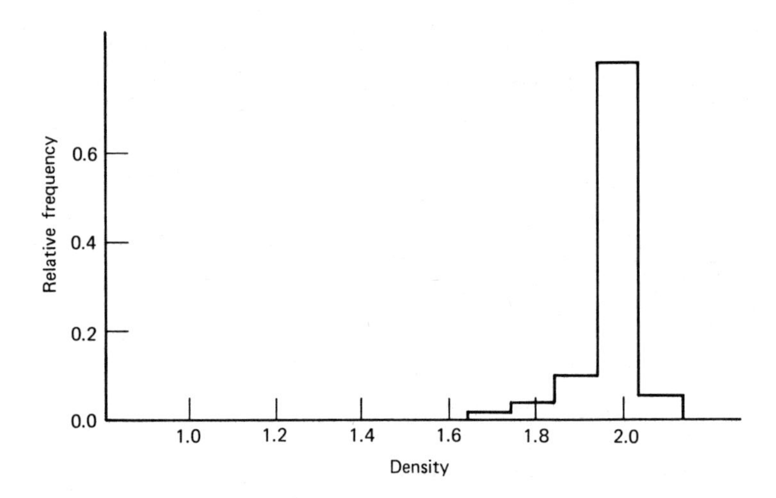

This plot ______ similar to a normal distribution. — is not
is, is not

28. The plot in item 27 is not much like the normal distribution because it is not symmetrical about the peak of the plot—it has a longer "tail" to the left than to the right. We call such a distribution "skewed." Something other than ______ is affecting the data. — chance

29. In this case, the reason that the distribution is skewed is simply that there is an upper limit to the density ("blackness") of a photographic paper—it cannot possibly go above a value of about 2.2. When there is a natural limit to a process, the distribution always tends to be ______ . — skewed

30. We expect a normal distribution only when the measurements have, for all practical purposes, no limits. The density of photographic film typically can vary from near zero to a large value—as much as 5 or 6. We would expect measurements in the middle of the range, say about 2, if they varied because of chance, to form a nearly ______ distribution. — normal

31. On the other hand, we would expect measurements near the low end of the range—near zero—to give a ______ distribution. — skewed

32. Scales of photoelectric meters often have numbers from 0 to 22. We would expect repeated measurements (with one meter and a standard test object) near the *ends* of the scale to be ______ . — skewed

33. We would expect similar measurements made near the *middle* of the scale to be nearly ______ . — normal

34. Another almost obvious departure from the normal distribution appeared when 75 light meters of the same brand were used to measure the light from a test object. The plot of the data was as follows:

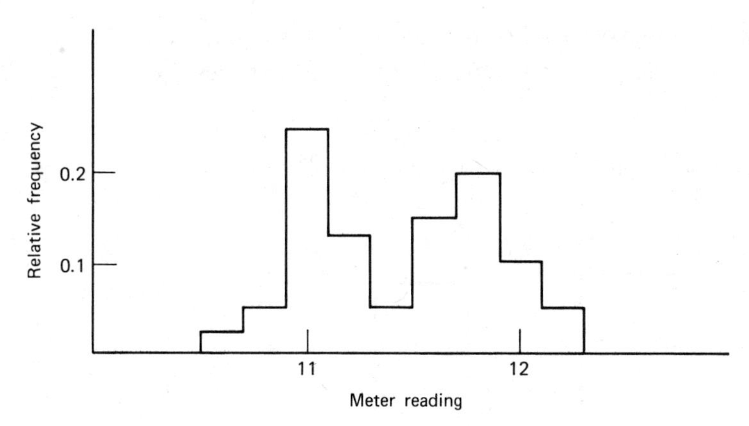

This plot ______ similar to the normal distribution shown in item 21. — is not
is, is not

35. The plot above is greatly different from the normal distribution because it does not have its highest point—its *mode*—in the center of the plot. Instead, the graph has two peaks—two modes—suggesting that the data came from two different sets-—two different *populations*. Examination of the meters showed that although they were the same brand, there were two different model numbers, and the suspicion that the samples came from two different populations was confirmed.

    If film samples, taken from *two* production lines, were measured for "speed," we would not be astonished to find that the plot showed two peaks, or two __________. — modes

36. A plot like that in item 34 is said to be *bimodal*. We suspect when we see a bimodal distribution that the sample data were taken from two __________. — populations

37. We have shown by the exceptional distributions in items 27 and 34 two of the important properties, or characteristics, of the normal distribution: it is symmetrical about the middle point of the data and has only a single __________. — mode

    The single mode (peak) always, for a normal distribution, lies at the center of the plot.

SELF-TEST ON INTRODUCTION TO VARIABILITY

Check your understanding of this section by answering the following questions. You find the correct answers after the questions.

1. Of the following numbers, which (if any) could, if correctly obtained, be exact: (a) a temperature of 25.2 C; (b) a light-meter reading of 9.3; (c) a count of 44 lenses in an inventory; (d) a shutter speed of 1/60 second.

2. A single shutter was tested for speed. Which of the following indicated more precise (careful) measurements: (a) the speed was always the same—1 second; (b) a series of tests gave 1.1, 1.0, 1.2, 0.9, and 1.0 seconds.

3. For a series of *80* trials in tossing a single coin, which of the frequency plots *(if any)* indicates a chance-controlled process?

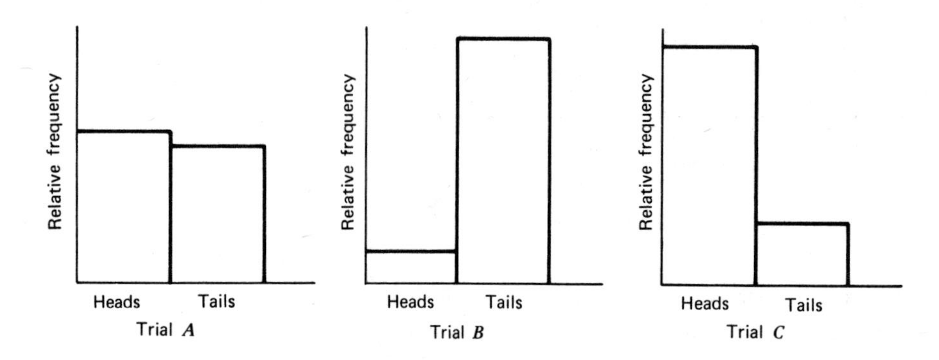

4. For a series of *8* trials in tossing a single coin, which of the frequency histograms *(if any)* indicates a chance-controlled process?

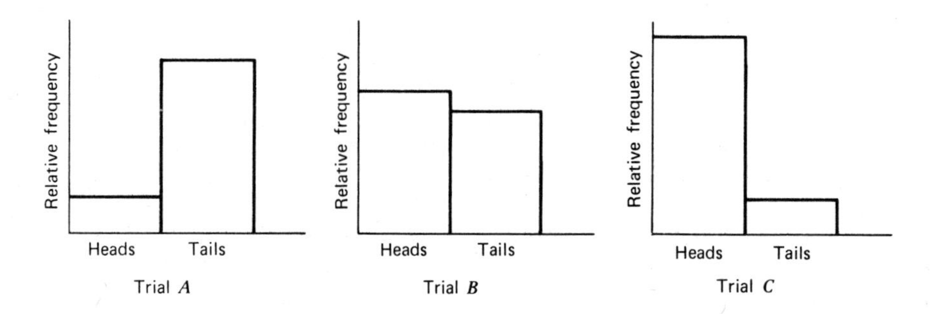

5. We expect the normal curve to be the underlying pattern for a set of data if:
   (a) we are dealing with ______________ values;
   measured, counted
   (b) the results are controlled by __________.

6. A skewed distribution results when the measured values have a(n) __________ or __________ limit.

7. A bimodal distribution results when the data come from two different __________.

8. Of the following distributions, which *(if any)* indicate(s) a process governed by chance? (The values represent measurements of the concentration of developing agent in 100 successive batches of mixed developer.)

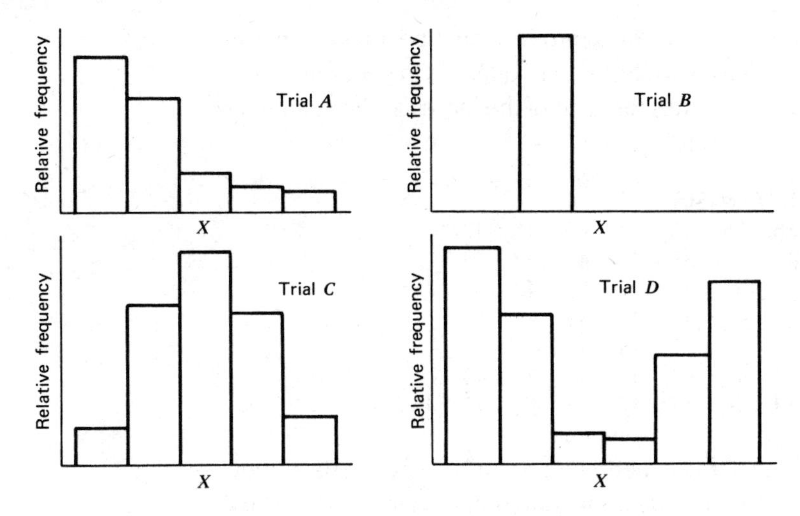

ANSWERS TO SELF-TEST ON INTRODUCTION TO VARIABILITY

In the parentheses that follow each answer you find the items in the preceding section that relate to the question and its answer.

1. (c) (Introduction).
2. (b) (items 4,5).
3. Trial *A* (items 7–10).
4. None (item 11—the sample size is too small).
5. (a) Measured; (b) chance (item 24).
6. Upper or lower (either order) (items 27–32).
7. Populations (items 34–36).
8. Trial *C* (items 24–25).

SECTION B
AVERAGES

We make repeated measurements on a process when we want to obtain useful information about the *population* (all possible data of a given kind). For reasons of cost and time, we practically never examine all possible members of the whole population. Instead, we must be content with a limited number of data points, and from this limited information try to make good judgments about the entire set. In some circumstances, to be specified in the first part of this section, we ccan indeed obtain good information about the population with only a few pieces of data.

In this section we refer often to the following table, which represents the results of this experiment: 100 lenses were taken from a production line making many thousands of similar lenses; each lens was tested for quality ("resolution"); the numbers in Table 2-1 are the measured values for each of the lenses as they came off the line:

TABLE 2-1 RESOLUTION OF TESTED LENSES

44	49	48	38	46	42	45	46	45	53	41	45	36	42		
47	46	47	45	44	43	54	43	41	39	48	47	52	44		
46	50	51	47	37	33	41	50	52	44	47	50	42	47		
38	30	46	44	53	46	43	47	47	47	45	38	45	48		
55	43	40	44	49	41	37	44	43	57	48	46	46	43		
45	41	45	42	43	52	43	42	41	43	43	35	42	45		
40	45	46	48	47	49	42	39	54	44	45	48	45	47	46	41

The data in the table came from a very large experiment; $n = 100$. Nevertheless, they are only a sample—only a part of the set of data that could have been obtained by testing *all* of the thousands of lenses.

The frequency histogram of the sample data is shown in the following graph.

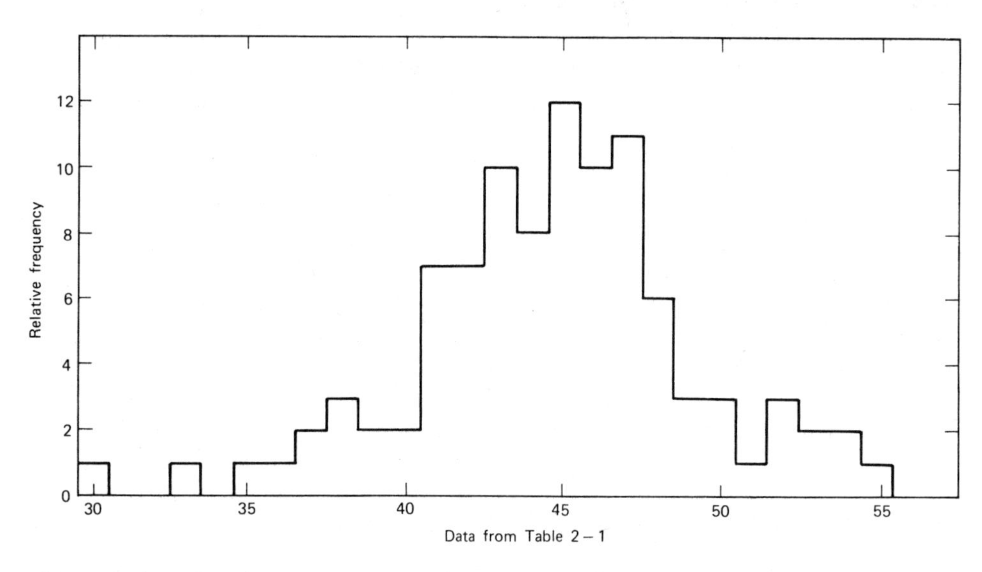

Although the plot shows some irregularities, you should agree that it resembles a normal distribution. The raw data in Table 2-1 as they stand are not very informative. The frequency histogram gives more insight into the process because it organizes the raw data to show the pattern. We can go beyond this to reduce the data to a small number of values that tell us nearly as much about the situation as the original data, in still more convenient form. In fact, for a normal distribution we need only *two* values to completely define the distribution. One of these, the *mean,* is discussed in this section.

WHAT YOU WILL LEARN

If you work carefully with the following material, you will know:

1. When the mean (average) is a useful number
2. How to find the mean of a set of data

Mean and Mode

DIRECTIONS: Cover about 2 inches of the right-hand margin of each of the following pages in turn with an opaque sheet of paper. Read carefully each of the numbered statements. On the cover sheet *write* your response to the blank left in each statement. Move the cover sheet down to expose the correct answer in the margin. Continue until you have finished the section. Work at a rate that is comfortable for you and no longer than you find pleasant. Begin when you wish.

1. In the preceding section you saw that the peak of a frequency distribution is called its ______ . — mode

2. For a normal distribution (hereafter, for short, ND) the value of the mode is the same as the value of the ordinary average of the data. We call the average the *mean*. For a *sample*, even a large one, the average will not exactly equal the mode but will be close to it. For example, the average (mean) of the data in Table 2-1 is 44.8. The mode (from the histogram on the previous page) is ______ . — 45

3. The mean value of 44.8 is very close to the mode, which is 45. The mode is the value that occurs most often in the entire set of data—the population. *If temperature measurements form a ND* and the mean value is 22 C, the value that occurs most often in the population is close to ______ C. — 22

4. If, on the other hand, the population is not a ND, the mean and the mode will *not* usually have the same value.
   Consider the distribution sketched as follows:

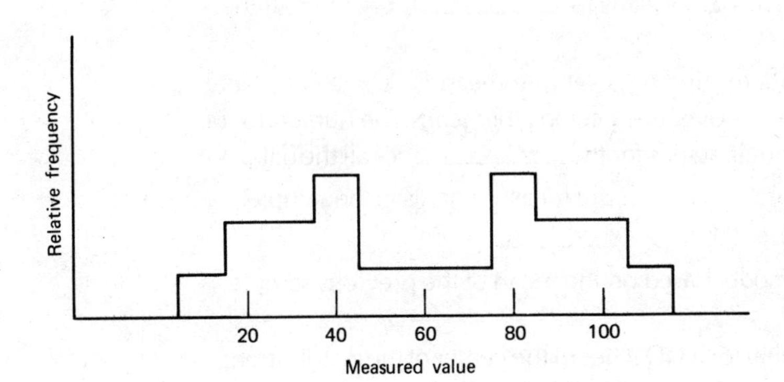

   This distribution ______ a ND. — is not
   is, is not

5. In fact this distribution is bimodal—it has two ______ . — modes

6. In the histogram in item 4, one of the modes is at the value ______ and the other at the value ______ . — 40, 80

7. The arithmetic average—the mean—would not be at either of the two modes. Instead, the mean would lie at the center of the whole distribution; it would have the value ______ . — 60

8. In item 4 is an example of a situation in which the mean is not of much value, because the distribution is not ______ . — normal

9. As another example of a case in which the mean is not very useful is shown by this skewed distribution:

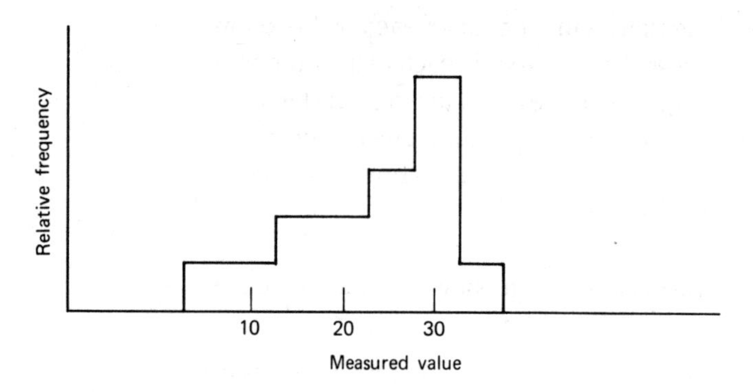

This distribution has its mode at the value ________ .

30

10. The mean of this distribution is *not* 30. It turns out in fact to be 23—well to the left of the mode. For a skewed distribution, the mean and the mode ________ have (do, do not) the same value.

do not

11. In general, the mean of a set of data is really useful for a ________ distribution.

normal

12. Whenever we average sample numbers, we will know (or assume) that the data come from a ________ distribution.

normal

13. You probably already know that to find the mean of a set of data we need to add them together and then to divide by *n*, the number of pieces of data. We use the symbol Σ to stand for the addition process. We use the symbol X_i to stand for each of the members of the data set. Thus ΣX_i means to ________ together all the measurements.

add

14. Using these symbols, the formula for finding a sample mean is: $\overline{X} = \Sigma X_i/n$. Note the symbol $\overline{X}$, read "X-bar." The line over the X stands for *mean*. The numerator of the fraction on the right of the formula stands for the ________ of all the data.

sum

15. The *n* in the formula stands for the ________ of measurements in the sample.

number

16. If for a sample ΣX is 155 and n is 10, $\overline{X}$ is ________ .

15.5

17. For a ND we would expect the mode, based on the mean of the previous sample, to lie near ________ .

15.5

18. Another property of the mean is that for a ND it lies in the center of the distribution; that is, half the values lie above the mean and half below it. This property arises from the symmetry of the pattern. For the data in Table 2-1, for example, the mean is close to 45. Forty-three of the data points have values greater than the mean; 45 of the data points have values less than the mean. (It would be remarkable if in a sample, even of $n = 100$, the data were *exactly* evenly divided.) Five samples of a slow film were measured for speed. The results were: 12, 10, 14, 12, 15. Use the formula $\overline{X}, = \Sigma X_i/n$ to find the mean. It is ________ .

12.6

19. Note that we carry out the value of the mean to just one more place than had the original X values. The reason is that averages of samples from a population differ *less* among themselves than do the individual pieces of data. Thus the $\overline{X}$ values, even for small values of *n*, are more reliably known than are the X values. We indicate this increased reliability by carrying out the computation of $\overline{X}$ to one more place. For example, the eighth column of the data in Table 2-1 gives $\Sigma X =$ 311 and $n = 7$. $\overline{X}$ is ________ .

44.4

20. In the same table, for the last full column, ΣX is 316, n is also 7, and $\overline{X}$ is ________ .

45.1

21. The two $\bar{X}$ values in items 19 and 20 are closely alike, much more so than the X values within the two columns. Also the sample means are very close to the mean of the whole sample of $n = 100$, which was 44.8. You will get similar results if you find the means of other samples in Table 2-1. For the first column, ΣX is 315. When we divide by n (7) the result comes out even. We still write the answer to one more place—45.0. We write the zero to show that the calculation of $\bar{X}$, is to the same number of places we would have used if it had not come out an even value. From another population a sample of $n = 3$ gave: 25, 30, 33. Use the formula for $\bar{X}$; it is __________ . 29.3

22. The answer to item 21 is 29.33333333 . . . ; we round off to the nearest tenth, dropping the 3s after that.
    For the data 7, 9, 8, 7, 8, 7, X is 46 and n is 6. $\bar{X}$ is __________ . 7.7

23. For item 22, dividing out gives 7.66666 . . . and we round up to the nearest tenth. A different sample: 8, 8, 7, 6. Before rounding off, $\bar{X}$ is __________ . 7.25

24. Here we have exactly 5 in the third place; we want to round off to the 10ths place. When the digit to be dropped (here, in the 100ths place) is exactly a "5," the rule is this: make the digit to be retained an *even* value. Here, rounding off 7.25 to the 10ths place, we simply drop the "5" and write the answer __________ . 7.2

25. If we were rounding off the value 7.35 instead, we would increase the "3" to a "4" and write the rounded-off value as __________ . The use of this rule means that in rounding off many values where there is a 5 in the place to be dropped, there will be no tendency to raise or lower the value consistently—about half the time we will raise it, and half the time we will lower it. 7.4

SELF-TEST ON AVERAGES

Check your understanding of this section by answering the following questions. The correct answers follow the questions.

1. For which (if any) of these distributions of measured data would the mean be a useful value?

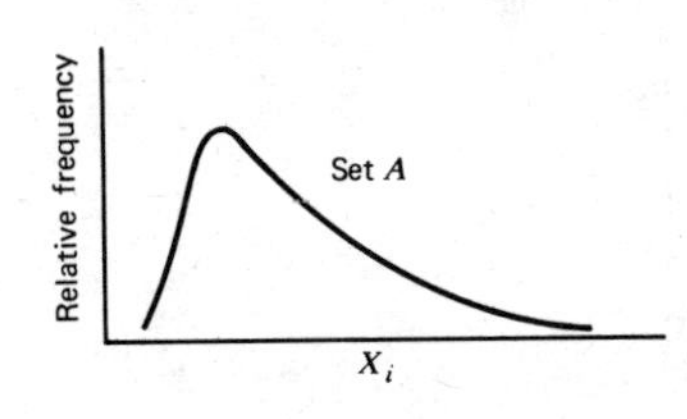

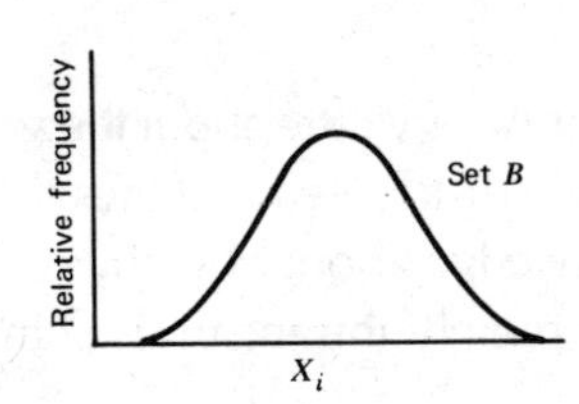

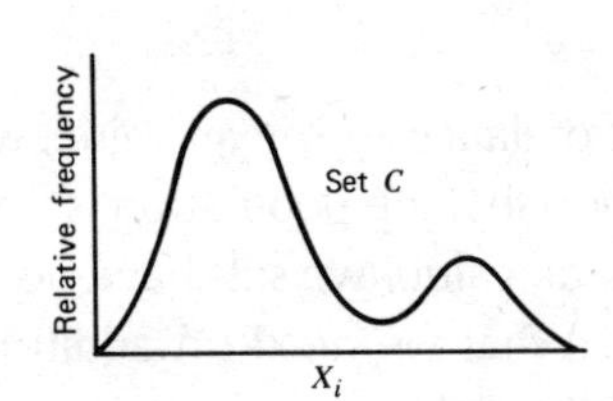

2. For a ND, the mean has the same value as the __________ .
3. For a ND, the position of the mean divides the data into two __________ parts.
4. In the expression:
   $\bar{X} = \Sigma X_i / n$
   $\bar{X}$ stands for the __________ ;
   Σ stands for the process of __________ ;
   X_i stands for each of the measured __________ ;
   n stands for the __________ of measurements.

5. A sample set of measurements is 5.4, 5.7, 5.8, 5.3.
   ΣX_i = ___________
   n = ___________
   $\overline{X}$ = ___________

6. Another set of sample measurements is 8.0, 8.3, 8.7, 7.9. Properly rounded off, $\overline{X}$ = ___________

7. In computing averages from samples, we assume that the samples are taken from a ___________ ___________ .
   (two words)

ANSWERS TO SELF-TEST ON AVERAGES

In the parentheses following each answer you find the numbers of the items in this section that relate to the question and its answer.

1. Set *B* (items 4, 9, 11).
2. Mode (items 2, 3).
3. Equal (item 12).
4. Mean (item 14); addition (item 13); values (or data) (item 13); number (items 13, 15).
5. 22.2; 4; 5.55 (items 13–15).
6. 8.22 (items 13–15; 24).
7. Normal distribution (item 12).

SECTION C
MEASURES OF VARIATION

If all we know about a set of data is the mean value, we know very little about the set. Even if we know in addition that the population is normally distributed and thus that the mean identifies the peak value, we still have no knowledge about how the data differ among themselves. What we need (in addition to merely the mean) is some measure of the spread of the data.

We can readily think of two NDs that have the same average, but different "widths," as in the two near-NDs sketched below. The data are for two different brands of film; the X values are speeds. Both brands would have nearly the same mean value, but brand *A* is more consistent—we would find less variation in the speed values from sheet to sheet or from roll to roll.

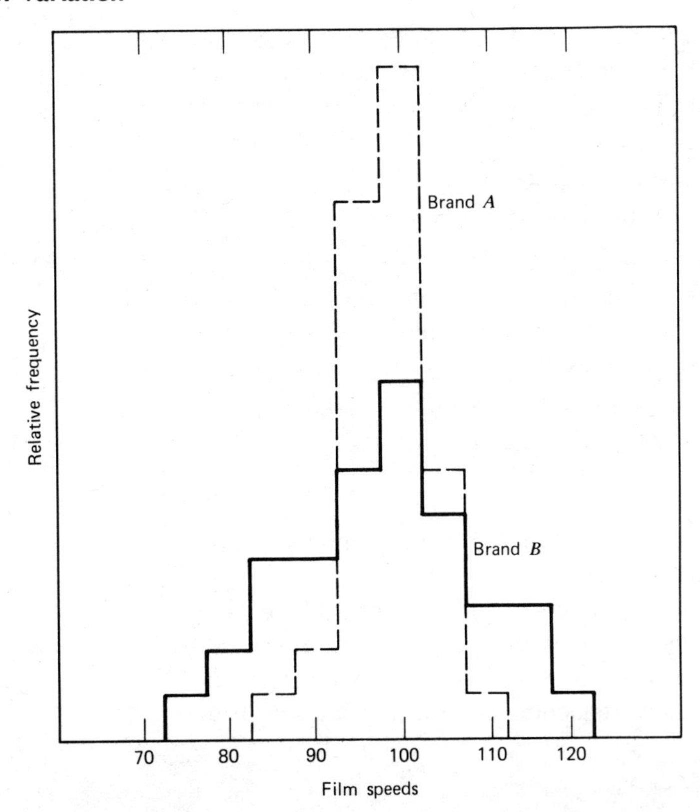

There are several measures of the extent of the variation that exists among the members of a set of data. In this section you will find a discussion of two such measures.

WHAT YOU WILL LEARN

If you work carefully with the following material, you will:

1. Be able to find the *range* of a set of data;
2. Know the advantages and disadvantages of the use of the range as a measure of variation;
3. Be able to find the *standard deviation* of a set of data;
4. Know the advantages and disadvantages of the standard deviation as a measure of variation.

DIRECTIONS: Cover about 2 inches of the right-hand margin of each of the following pages with an opaque sheet of paper. Read carefully each of the numbered statements. *Write* your response to the blank left in each statement on the cover sheet. Continue until you have finished the section.

1. The simplest measure of the spread of a set of data is called the "range." It is merely the extreme difference in the values, obtained by subtracting the smallest from the largest. Refer to the immediately preceding histograms; for brand *B* the largest value of the speed is __________ . — 120
2. For brand *B* the smallest-speed value is __________ . — 75
3. The range is the difference between the two *X* values, and is __________ . — 45
4. For brand *A* the range is __________ . — 25
5. The range is an attractive measure of variation because it is very easily found. Refer to the data at the beginning of Section B. From the table, or more easily from the histogram, you find that the range of resolution values is __________ . — 27
6. Like many things that come easily, the range has only limited usefulness. One reason is that the range is based on only ________ (number) of all the 100 pieces of data accumulated in the lens tests. — two
7. Thus the size of the range is determined solely by the largest and smallest values in the set, and ignores all the rest of the data.

   Furthermore, when *n* is large, it is likely that we will find at least one very large and one very small *X* value, even though the rest of the data may cluster closely about the mean. One reason for the preceding statement is that the "tails" of a normal distribution *never* fall to zero frequency, as shown in the sketch that follows.

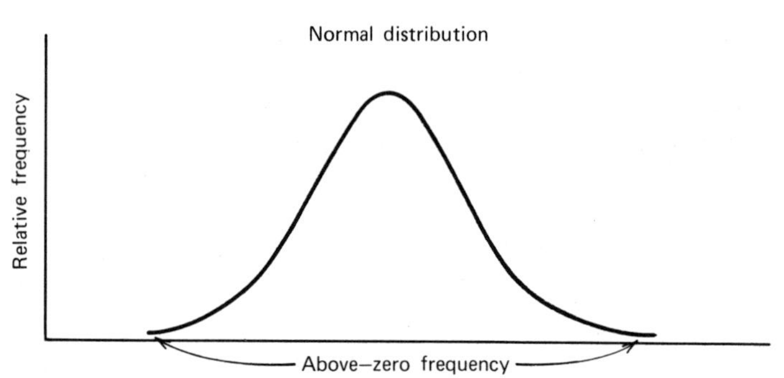

   There is therefore always a real (if small) chance of getting extreme values from a population that is ND, and this causes the range to be exaggeratedly __________ . — large
8. Thus the range is an unreal estimation of the variability, especially when the value of *n* is large.

   Specifications for processes often (as they invariably should) give the *tolerances,* that is, the allowable variation in the desired value. A common example is developing temperature, where the instructions may say to process at 30 C "±½." The total range allowed by this statement is __________ . — 1 C
9. Such a statement of the tolerance *is not* a good one for this reason: even in a process that is ND and centered on the aim value of 30 C, some extreme temperature values will be found if many measurements are made. The truth might well be that by far *most* of the temperatures met the specifications. If in a given set of many temperature measurements you found a few outside the tolerances, to condemn the temperature-control system because of these few readings would probably be __________ (right, wrong). — wrong

10. It is not good practice to give tolerances as ± values, and it is not good practice to judge processes on the basis of the extreme values, because the range is not a good measure of variation if n is __________ . large

11. Since the tails of a ND never fall to zero, to give a range that will include *all* possible X values is __________. impossible
possible, impossible

For the reasons indicated in items 6–11, a better measure of variability is needed, namely, one that (a) makes use of all the data, not just two, (b) is useful for large as well as small sample sizes, and (c) more effectively describes the variation. This measure is called the "standard deviation." It is free of the defects of the range, but is primarily applicable to the *ND*.

12. The standard deviation is like a "root-mean-square." It is harder to find than the range but is worth the effort because it is so useful. One method of finding the standard deviation is by the following steps. We will use simple data for illustration: the X values are 4, 8, and 6. STEP 1. Find $\overline{X}$: __________ . 6.0

13. STEP 2. FInd and list all of the differences between $\overline{X}$ and the original data set, ignoring the sign of the difference. In order, the differences are __________, __________, and __________ . 2.0, 2.0, 0.0

14. STEP 3. Square all of the differences. In order, the squares of the differences are: __________, __________ and __________ . 4.0, 4.0. 0.0

15. STEP 4. Sum the squared values; the result is __________ . 8.0

16. STEP 5. Divide this sum by (n-1); here divide by __________ . 2

17. STEP 6. Finally take the square root of the quotient. The answer is __________. 2.0

NOTE: In these calculations, we work to one more decimal place than in the original data; we write the standard deviation to the same number of *decimal places* as we write the mean.

18. These six steps can be put into this formula: $s = \sqrt{\Sigma(\overline{X}-X_i)^2\,(n-1)}$. Here only "s" is a new symbol; it stands for the standard deviation found from a *sample*. $\overline{X}$ stands for the sample __________ . mean

19. X_i stands for each of the pieces of data in the sample. n stands for the sample __________ . size

20. Σ is the sign meaning to __________ what follows the sign. add

21. Follow the same six steps to find s for a different sample. It is helpful to work systematically using a table form. We have put the Xi values into the first column of the table:

X_i
60
70
70
80
90

STEP 1. Find $\overline{X}$; it is __________ . 74

22. STEP 2. In the next column of the table, list all the differences between $\overline{X}$, and each X_i:

X_i	$(\overline{X} - X_i)$	
60	________	14
70	________	4
70	________	4
80	________	6
90	________	16

23. STEP 3. Put the squares of the differences in the next column:

X_i	$(\overline{X} - X_i)$	$(\overline{X} - X_i)^2$	
60	14	________	196
70	4	________	16
70	4	________	16
80	6	________	36
90	16	________	256

24. STEP 4. Find the sum of the numbers just found; it is ________ . 520
25. Divide 520 by (n-1), that is, by 4, and get ________ . 130
26. Take the square root of 130; it is ________ . 11

The standard deviation of the sample is 11, written to the units place, since the original data were given to the nearest 10 only.

We leave for the time being the computation of the standard deviation and treat in the following the meaning of s. It has many applications, but *only* if the following requirements are met:

(a) The data must come from a normal distribution;
(b) The sample must be representative of that population;
(c) The sample must be of reasonable size, preferably at least about 30. (Fewer will do, at the risk of less assurance about the inferences to be made.)

If these three assumptions are correct, the following drawing represents the model, namely, the population distribution:

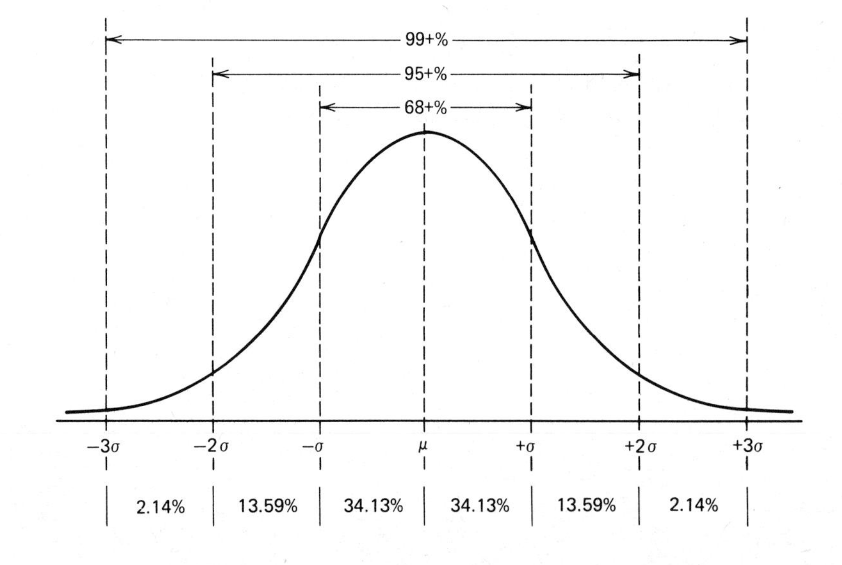

27. The graph represents a ____________ distribution. — normal

28. Note the symbolism used here μ (Greek "mu"), since it is the X value for the peak, stands for the population ____________ . — mean

29. If the sample is properly chosen, μ (the *population* mean) is approximated by the *sample* mean, symbolized by ____________ . — $\bar{X}$

30. $\bar{X}$ is closer to μ as the sample size increases. In the graph above, σ (Greek "sigma") stands for the population standard deviation. Again, if the sample is correctly selected, the population standard deviation (σ) is approximated by the *sample* standard deviation, symbolized by ____________ . — s

31. We take a sample, and find its $\bar{X}$, and s values when we are interested in estimating the corresponding population values, ____________ and ____________ . — μ, σ

32. If in fact the sample values of $\bar{X}$ and s are good estimates of the population values μ and σ, then the percentage values on the graph on the previous page can give us great insight into the situation. Note that the horizontal axis is scaled in units of the population standard deviation σ. Let us assume that μ is 100 and σ is 10. We could then place numerical values on the horizontal axis: we put 100 at μ; 100 + 10 (110) at + σ; and at +3 σ we put ____________ . — 130

33. Similarly, at - σ we put 100-10 (90); and at -2 σ we put ____________ . — 80

34. We now have the graph with numerical values shown below:

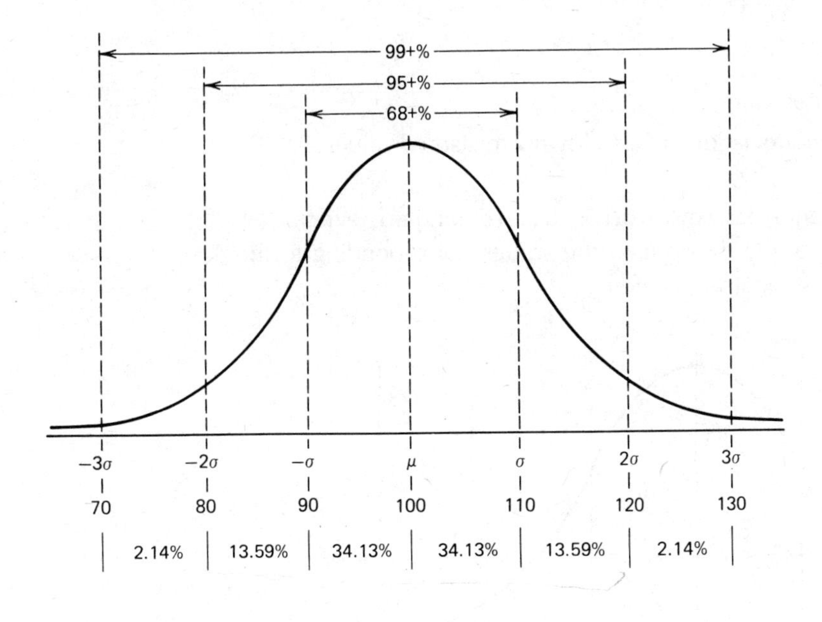

The percentage values above and below the graph show the fraction of the population lying in the indicated areas of the curve. Thus a little over 34% of the population data lie between 100 and 110. Only about ____________% lie between 120 and 130. — 2

35. About 68% of the data fall between 90 and 110 (i.e., between $\mu \pm \sigma$). Between 80 and 120 we find about ____________% of the population. — 95

36. Almost all the population falls between $\mu \pm 3\,\sigma$; the percentage is ____________ — 99+

SELF-TEST ON MEASURES OF VARIATION

Check your understanding of the previous section by answering the following questions. The correct answers follow the questions.

1. For the following set of measured sample data, what is the value of the range? 7, 8, 4, 10, 6.
2. There are two reasons that the range is a relatively poor measure of variation. One reason is that the computation of the range makes use of only ________ pieces of data from the entire set.

   The other reason is that for large sets of numbers the range is likely to be too ________ .
3. Find the mean of the values in (1). It is ________ .
4. Using the tabulation below, find the sample standard deviation:

X_i	$(\overline{X} - X_i)$	$(\overline{X} - X_i)^2$
7	________	________
8	________	________
4	________	________
10	________	________
6	________	________

   The sum of the values in the right-hand column is ________ .
   n-1 is ________
   The sum divided by (n-1) is ________ .
   The square root of the previous value is ________ .
5. Find the range and the standard deviation of the following measured values: 21, 32, 24, 19.
6. A normally distributed population has a mean (μ) of 50 and a standard deviation (σ) of 5. Enter below the graph of the distribution the values corresponding to the whole-number values of the standard deviation.

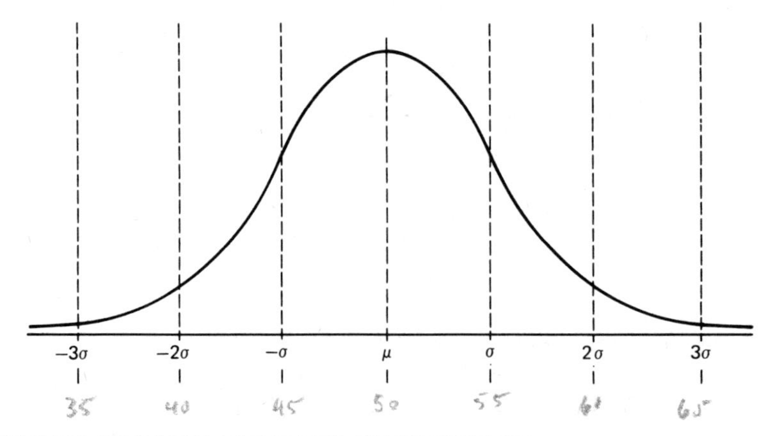

ANSWERS TO SELF-TEST MEASURES OF VARIATION

In the parentheses following each answer you find the numbers of the items in this section that relate to the question and its answer.

1. 6, that is, 10 -4 (items 1–4).
2. Two; large (items 7 and 8).
3. 7.0 (item 12).
4.

X_i	$(\bar{X} - X_i)$	$(\bar{X} - X_i)^2$
7	0	0
8	1	1
4	3	9
10	3	9
6	1	1

Sum = 20; $(n-1) = 4$; $20 \div 4 = 5$, and the square root of 5 is *2.2* (items 21–26).

5. The range is 5 and the standard deviation is 5.7 (items 1–4, 21–26).
6. The values in order from left to right are: 35, 40, 45, 50, 55, 60, 65.

SECTION D APPLICATIONS OF THE STANDARD DEVIATION

In the preceding section you saw that of the different measures of variation, the standard deviation is by far the most useful. Knowing the population σ, and knowing that it is normally distributed, we can then find reliable answers to questions like these:

1. How often can we expect a process to meet the requirements? Example: suppose that we obtain good color transparencies only when the development temperature lies between 74 F and 76 F; we wish to know whether or not a given temperature-control system will be satisfactory. We certainly need to know the standard deviation of the process temperatures in order to know that the process will do what it is supposed to.
2. Do new data indicate that a process has changed? Example: a light meter when tested with a standard source has a long history of readings that average 11.5; a later test gives a reading of 12.5. We cannot suppose that this new reading indicates a change in the meter performance unless we know the variations in the data that made up the original average.
3. Are two similar processes enough alike so that they can be used interchangeably? Example: one light meter gives average readings of 11.5 and another, an average of 12.5. Our judgment about whether the meters are really different depends on our knowledge of the variability of each of them. (Perhaps the first meter often gave readings near 12.5.)

The three types of question mentioned above are basic and of wide practical importance. Methods for handling them are discussed in books on applied statistics. Here we consider the first type of problem—"conformance to specifications."

WHAT YOU WILL LEARN

If you work carefully with this material, you will be able to:

1. Determine the necessary process standard deviation in order to meet stated requirements;
2. When the process standard deviation is known, find out how often a given process will be able to meet requirements.

DIRECTIONS. Cover about 2 inches of the right-hand margin of each of the following pages in turn with an opaque sheet of paper. Read carefully each of the numbered statements. *Write* your response to the blank left in each statement on the cover sheet. Continue until you have finished the section.

Please refer to the temperature example in the introduction to this section, the paragraph numbered 1.

1. The permissible temperature range is ___________ F. — 2
2. No doubt the desired average (mean) temperature is midway between the two allowable extreme values; thus the desired (*aim*) processing temperature is ___________ F. — 75
3. Thus the *specifications* for this process would read: "75 F ± 1." The aim is 75 F, and the *tolerance* is the allowable variation on either side of the aim, that is, ___________ F. — 1
4. We will now consider the extent to which a possible process would meet these specifications. Suppose that the actual variations in process temperature are represented by a ND and thus caused only by ___________ . — chance
5. The best possible case would be one in which the true mean temperature was exactly ___________ F. — 75
6. (In fact, the actual mean temperature would be never exactly equal 75 F, so we are imagining an ideal that is never attained in practice.) The sketch that follows represents in part this ideal case:

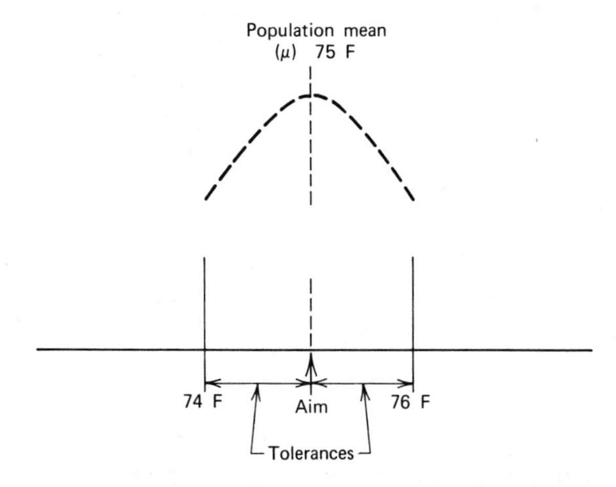

We have shown the peak (mean) of a ND lying midway between the two specification limits of 74 F and 76 F. The true process average (μ) is identical with the aim value of ___________ F. — 75

7. As yet we have not completed the sketch. Missing (as suggested by the dotted lines) is the rest of the ND curve. To finish the sketch, we need to know the width

(or spread) of the actual process data, as measured by its __________ __________ . — standard deviation

8. Now suppose that this temperature-control system gives a standard deviation equal to the tolerance, and therefore that the value of σ is __________ F. — 1
9. We can now finish the sketch of the situation, completing the ND curve:

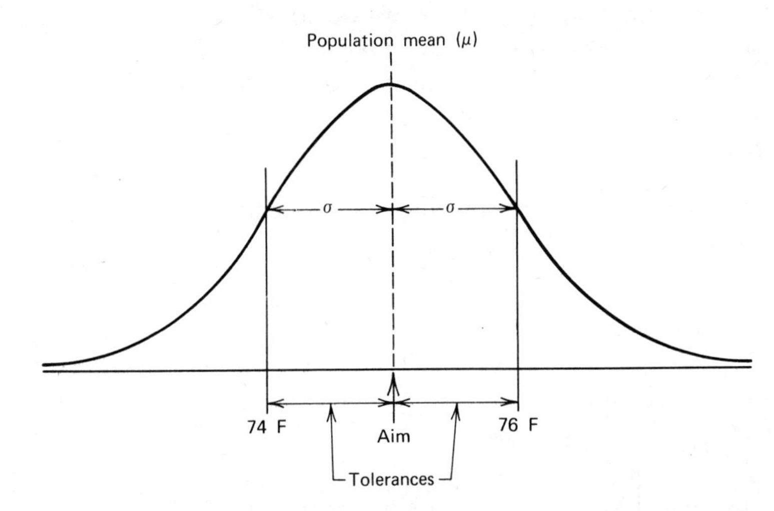

The question now is how often would this process meet the requirements; to answer this, we repeat from Section C the marked areas of a ND:

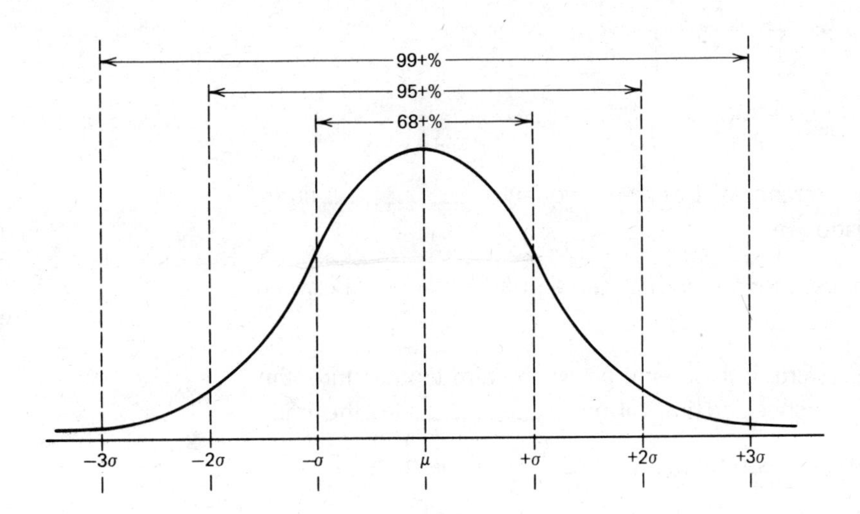

Note over the curve just above the marked area between $-\sigma$ and $+\sigma$. It is __________% . — 68

10. Thus for a normally distributed set of data, 68% of the values will lie within this $\pm 1\sigma$ range. Since we are assuming that for this process this range is also the tolerance range, we know that this process will control the temperature to the desired values only __________% of the time. — 68
11. The difference between 68% and 100% is the percentage of the time that this temperature control process will *not* meet the specifications. It is __________% . Thus if the process standard deviation equals the tolerance, the temperature will — 32

be incorrect almost a third of the time, even if the process is exactly centered on the aim. This process is certainly not adequate to do the job.

12. Since the *chance* variations in this process are such as to give temperatures outside the specification limits, it would be *impossible* by adjusting the average temperature level to obtain the correct temperatures. If we found a temperature of 77 F, for example, and changed the setting so as to lower the mean value, we would often be wrong in doing so since the *average* temperature might well be correct.
   To meet the specifications of 75 F ± 1, we need a better system, that is, one with a smaller ____________ ____________ . (two words)

standard deviation

13. Let us now consider a different temperature-control system, one with a standard deviation of only ½ degree. The situation now is this:

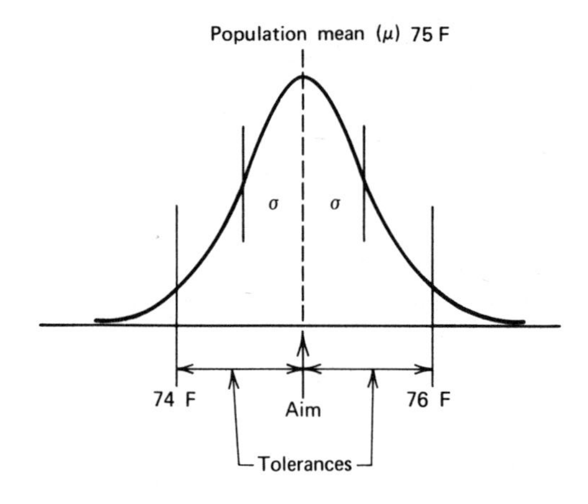

For this better process, the tolerance of 1 degree is equal to ____________ times the process standard deviation σ.

two

14. Refer again to the marked areas for the normal curve in item 9 above. Between $\pm\ 2\ \sigma$ you find an area of ____________%.

95+

15. Therefore for this second system, if it is centered on the aim temperature, the temperatures will be within the specified limits about ____________% of the time.

95

16. The temperature will be wrong less than ____________% of the time.

5

17. The second and better temperature-control system would give 95% conformance to specifications. Furthermore, since the curve area is about 68% between $\pm\ 1\sigma$, over two-thirds of the time the temperature would be within ____________ degree of the aim.

½

18. Perhaps 5% out-of-specifications temperatures will be satisfactory. If not, a still better process is needed—one with still smaller standard deviation. Such an improved process is sketched here.

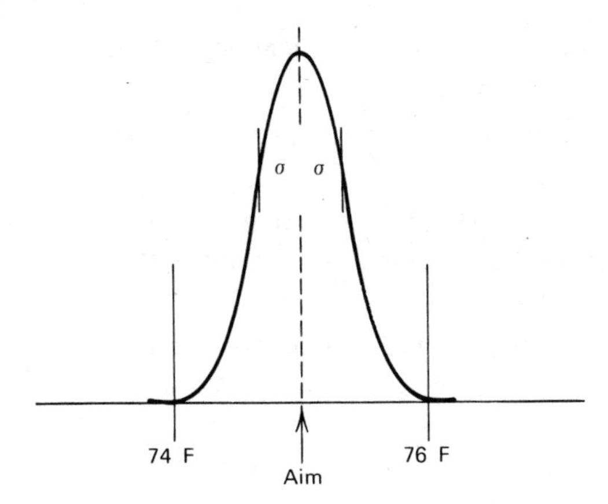

For this third process, σ is 1/3 degree F. Since the tolerance is equal to 3σ, this system will meet the specifications (if it is centered on the aim) over ___________% of the time. — 99

19. For this process, the temperatures would be incorrect less than ___________% of the time. — 1

More precisely, the process with σ equal to one-third the tolerance will meet the specifications (if it is centered on the aim) 99.72% of the time and will fail to meet the specifications only 0.27% of the time—less than three times in 1000.

For this process, operating as we have assumed, it would be a very rare event to have the temperature rise as high as 77 F, unless there had been a change in the process mean, so that the entire set of temperatures had risen. Therefore it would be reasonable to conclude (if a value of 77 F were found) that something had caused a change in the mean temperature, and then to adjust the process level.

20. There are innumerable applications of the use of the standard deviation to relate actual process performance to specifications.

It is wise, for example, to check the operation of a densitometer—an instrument for measuring the density of a photographic image. For this purpose we use a test object having a known density, say 0.78. We agree to accept any reading lying between 0.74 and 0.82. Our tolerance is thus ± ___________ — 0.04

21. We will accept any value lying within 0.04 of the aim, which is ___________. — 0.78

22. If many readings with a given densitometer give a near-normal distribution centered on 0.78, we know that the measurement average is correct. Suppose the measurement standard deviation is 0.02. The tolerance is equal to ___________ times the standard deviation. — 2

23. We know that the area of the normal curve between $\pm 2\sigma$ is ___________%. — 95+

24. Thus, we know that if the average of the densitometer does not change, it will meet the requirements ___________% of the time. — 95+

25 This instrument would, even if it does not change in average performance, give unacceptable readings about ___________% of the time. — 5

26. If we tightened up our specifications, so that we would accept readings only between 0.76 and 0.80, our tolerance is now ± ___________. — 0.02

27. With this smaller tolerance, the instrument would meet the specifications only for the area lying between $\pm 1\sigma$, which is ____________% . 68+

28. With this reduced tolerance, the densitometer would give unacceptable readings about ____________% of the time. 32

29. If, on the other hand, we could accept a greater tolerance, as much as 0.06 on either side of the aim, we would expect this densitometer to give acceptable readings more than ____________% of the time. 99

30. You see that if we express the tolerance in multiples of the actual process standard deviation, we can readily determine the conformance to specifications from the areas for the normal curve.

    Assume that specifications for camera shutters supposed to operate at 1 second have a tolerance of 0.12 second from the aim. Suppose a group of tested shutters give times that are ND and have a mean value of 1.00 second. To have 95% conformance to specifications, the shuttter standard deviation would need to be ____________ . 0.06

31. If the tolerance is equal to 2σ, we found that 95% of the values for a ND would lie between $\pm 2\sigma$. If the shutters had less variability, so that the standard deviation was only 0.04, we would have over ____________% conformance to specifications. 99

32. If, on the other hand, the shutters were more variable, and had a standard deviation of 0.12, we would have only ____________% conformance to specifications. 68

SELF-TEST ON APPLICATIONS OF THE STANDARD DEVIATION

Check your understanding of this section by answering the following questions. The correct answers follow.

1. For a given application, the speed of a film must be 50 ± 10. The aim value is ____________ and the tolerance is ± ____________.
2. Same situation as in item 1. If the average film speed is in fact 50 and the standard deviation of the speed values is 10, the process will meet the requirements ____________% of the time. The process will fail to meet the specifications ____________% of the time.
3. Same situation as in item 1. To meet the specifications 95% of the time, the standard deviation of speed values would need to be ____________ .
4. Same situation as in item 1. To meet the specifications over 99% of the time, the standard deviation of speed values would need to be about ____________ .
5. In problems like those above, we are assuming that the data, when plotted, would form a ____________ ____________ .
   (two words)
6. Of the following sketches, which represents a situation in which the specifications would be met almost all the time? In each case, the vertical bars represent the specification limits, and the curves represent the actual distribution of the data.

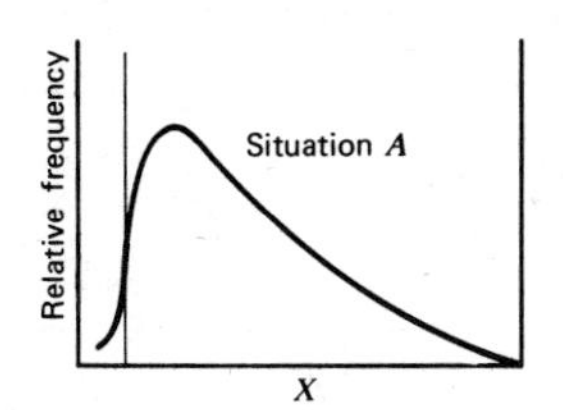

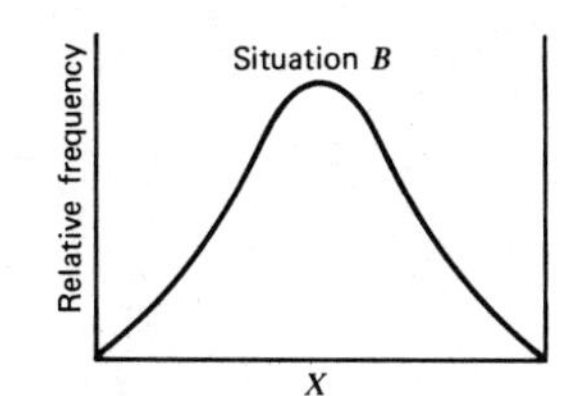

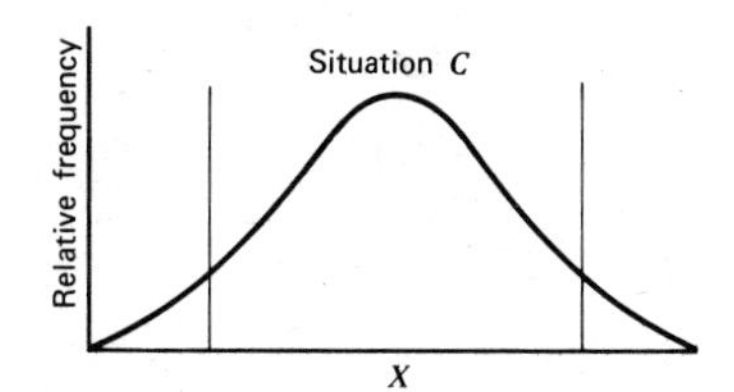

ANSWERS TO SELF-TEST ON APPLICATIONS OF THE STANDARD DEVIATION

In the parentheses following each answer you find the numbers of the items in this section that relate to the question and its answer.

1. 50; 10 (items 2, 3).
2. About 68%; about 32% (items 9–11).
3. 5 (items 13–17).
4. 3.3 (items 18, 19).
5. Normal distribution (items 4, 22).
6. Situation *B* (items 18, 19). Note that only this distribution, of the three given, is the near-normal and has a sufficiently small spread.

CHAPTER 2

Introduction to the Characteristic Curve

In manufacturer's and other literature, it is common to find a graph called the *D*–log *H*, or "characteristic" curve, which is in a useful sense descriptive of the response of a photographic material (film or paper) as this response is affected by what is done to the material in the process. Usually a series of such graphs displays the effects of altered development time, as in the example below. The plot shows on the vertical axis the response of the material; it is labeled *density*. It shows on the horizontal axis measures of the *exposure*, that is, the quantity of light that fell on the sample per unit area.

Such a set of graphs is obtained from an experiment in which samples of film are given known amounts of light (exposures), usually with an apparatus called

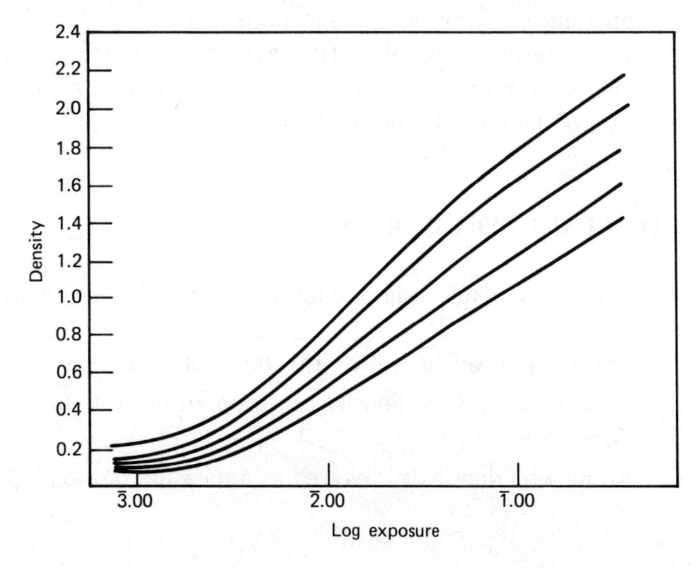

a *sensitometer*. As the sketch below shows, a sensitometer comprises a stable light source, an accurate shutter, and a "step tablet" that is basically a set of side-by-side neutral filters to change the light level at different places on the sample (see Appendix D for a discussion of the concept of exposure and Appendix B for alternative test methods).

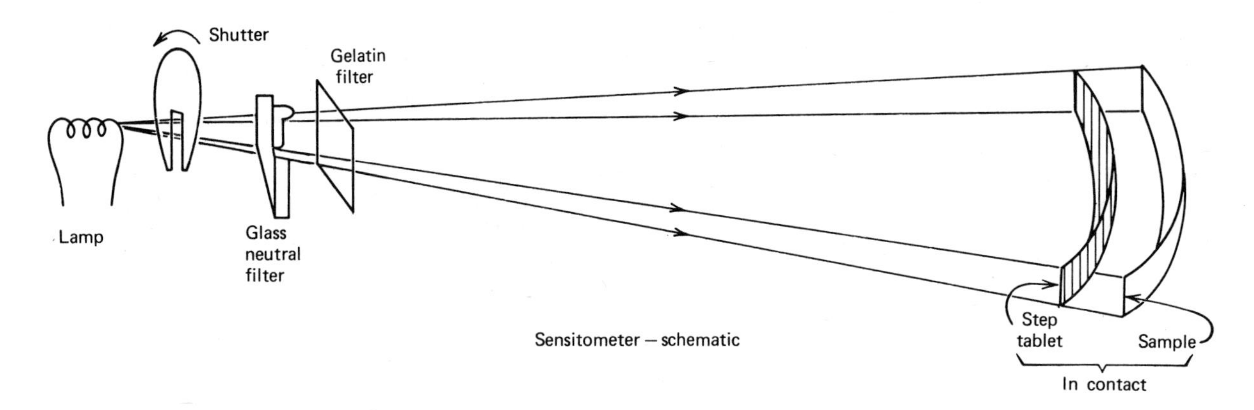

Sensitometer – schematic

After the samples are processed, they are measured with an instrument called a *densitometer*, which is a sensitive light meter capable of measuring the light passing through a small area of the sample film. From the resulting data (known exposures and known densities) a plot is made for each different development time, giving a "family" of curves like those above.

For reasons explained in this section, both the input exposures and the sample densities are usually expressed in logarithms. In what follows, it is assumed that the reader is familiar with the basic concepts of logarithms, especially Appendix Sections A1 and A3.

We begin with the concepts that lead to the use of density as a measure of the photographic image—the photographic response.

A negative from a conventional (silver halide) photographic process usually consists of many different areas (tones), each of which contains a different amount of finely divided silver. The amount of silver in each area will have been determined by the light the sample received in that area and also by the development conditions (time, temperature, chemistry, and agitation).

When the negative is placed in the printer, each different area of the negative allows a different amount of light to pass through, and thus affects the light received by the corresponding area of the print paper.

WHAT YOU WILL LEARN

If you work carefully with the following material, you will:

1. Know the meaning of the terms transmittance, opacity, and density;
2. Be able to find the density of a sample from its transmittance or opacity, and conversely;
3. Know why density is preferred as a measure of the photographic image.

DIRECTIONS. Cover about 2 inches of the right-hand margin of each of the following pages with an opaque sheet of paper. Read carefully the first numbered statement. *Write* the word or phrase that you believe correctly completes the statement. Move the cover sheet down to reveal the correct answer in the margin. Continue in this way until you have finished the section. Work at a rate that is comfortable for you, and no longer than you find pleasant. Begin when you wish.

1. The process by which light passes through a film negative is called "transmission." The fractional part of the original light that penetrates the film sample is called the "transmittance." If a portion of a negative receives 100 units of light and passes 50 units, the fractional part of the light that passes through the negative is ________ .

   ½, or 0.5

2. If the negative receives 100 units of light and transmits 50 units, its *transmittance* is ½, more usually written as the decimal fraction 0.5. If another part of the negative receives the same 100 units of light but transmits only 10 units, its transmittance is ________.
   (decimal fraction)

   0.1

3. A sample receives 100 units of light and transmits 1 unit. Its transmittance is ________ .

   0.01

4. If a sample receives 250 units of light and transmits 50, its transmittance is ________ .

   0.2

5. Transmittance is always a fraction, less than 1, because no real sample transmits the light completely. Nearly clear film base, for example, may receive 100 units of light and transmit as much as 90. Its transmittance is ________ .

   0.90

6. A completely opaque sample would transmit no light. If it received 100 units of light and transmitted none, its transmittance would be ________ .

   0

7. A negative is never completely opaque. Some negative materials can produce so much silver that a sample receiving 1000 units of light would transmit only 0.1 unit, and therefore have a transmittance of only ________ .

   0.0001

8. Thus the transmittance of a negative may range from nearly (but not quite) 1 for an almost clear part of the negative to nearly (but not quite) 0. As the amount of silver in the negative goes up, the transmittance goes down. Thus the converse of transmittance is more directly related to the amount of silver. The converse of transmittance (i.e., the "light-stopping" ability of the sample) is called "opacity." The opacity is found by dividing the transmittance into 1. If the sample transmittance is 0.5, the opacity is 1/0.5, or ________ .

   2

9. An opacity of 2 means that the sample receives twice as much light as it transmits (allows to pass). If the transmittance of a sample is 0.1, its opacity is 1/0.1, or ________ .

   10

10. The operation of dividing a number into 1 is called finding the *reciprocal* of the number. Thus the opacity of a sample is the reciprocal of the transmittance of the sample. Conversely, the transmittance is the reciprocal of the opacity. To find the reciprocal of a number, make it the denominator of a fraction with numerator 1. If the transmittance is 0.4, the reciprocal is 1/0.4, which on dividing out gives ________ .

    2.5

11. Conversely, if the opacity is 100, we find the transmittance by finding the

reciprocal of 100, that is, 1/100. If the opacity is 200, the transcittance is ______. (decimal fraction) — 0.005

12. Following the reciprocal rule, if the transmittance is 0.2, the opacity is ______ . — 5
13. If the opacity is 5, that means that five times as much light strikes the sample as passes through it. If the sample transmittance is 0.02, the opacity is ______. — 50
14. For a transmittance of nearly 1, the opacity is also nearly ______ . — 1
15. For a transmittance of nearly 0, the opacity will be a very large number. For example, if the transmittance is 0.001, the opacity is ______ . — 1000
16. Thus for any sample the opacity will be at least 1, and may be an indefinitely large number. You may find either the transmittance or the opacity from the light values. If 75 units of light fall on the sample and 25 pass through, the fraction 25/75 gives you the *transmittance,* and a decimal fraction of about ______. — 0.33
17. If you use the same numbers but write 75/25, you are finding the *opacity* directly; it would be ______ . — 3.0
18. Thus both transmittance and opacity involve a ratio of the same light values but in inverse order. If the sample receives 100 units of light and allows 5 units to pass through, the transmittance is ______ and the opacity is ______. — 0.05, 20
19. Although opacity increases as the amount of silver increases in the image, it does *not* do so *proportionally*. In part for this reason, *density* is more often used as the measure of the image. Density is the *logarithm* of the opacity. If the opacity is 10, the density is ______ . — 1

    (See Appendix Section A1, items 13–19.)
20. Some negative materials have a maximum opacity of about 10,000. The maximum density of such materials would be about ______ . — 4
21. The opacity of a negative is about 2 in the shadow areas. The density of such areas is about ______ . — 0.3

    (See Appendix Section A3, item 12.)
22. A face tone in a portrait negative may have an opacity of about 8 and thus a density of about ______ . — 0.9
23. If we know the density of a sample, we may find its opacity and transmittance. If the density is 1.3, the opacity is the antilog of the density, or ______ . — 20
24. If the opacity is 20, the transmittance is the reciprocal of 20, or ______. (decimal fraction) — 0.05
25. If the density of a sample is 2, the opacity is ______ and the transmittance is ______ . — 100, 0.01
26. To a good approximation, when properly measured, density is directly and nearly proportionally related to the amount of silver in a given area of a negative. Thus if two spots on a negative have densities of 0.60 and 1.20, because the larger density is twice the smaller, there will be in the spot having the larger density ______ as much silver as in the other spot. — twice
27. If two areas of a negative have densities of 0.30 and 1.50, the denser area will have about ______ times as much silver as the other area. — five
28. The amount of silver is the *effect* on the photographic material caused by the

photographic process. Because of the simple relationship between density and the amount of silver, it is logical to use *density* as the measure of the photographic effect. In part for this reason, it is customary to find density as the value on the vertical axis of the characteristic curve, as on the first page of this section. Refer to that set of curves, and note that as you read from left to right, every curve shows an increase in density, and thus an increase in the amount of __________. — silver

29. Suppose that three areas of a negative have densities of 0.30, 0.60, and 0.90. Because the densities increase uniformly from area to area, it would be reasonable to say that there is also a uniform increase in the amount of __________ in these areas. — silver
30. The opacities (the antilogs) of the same density values are, in order, __________, __________, and __________. — 2, 4, 8
31. Note that the opacities do not increase uniformly, in the arithmetic sense, since the second is 2 more than the first, but the third is __________ more than the second. — 4
32. Thus a uniform increase in density *is* associated with a uniform increase in the amount of silver, that is, in the __________ __________ (two words). — photographic effect
33. A uniform increase in the amount of silver, on the other hand, does *not* produce a uniform increase in the opacity. We therefore prefer __________ as a measure of the photographic effect. — density

We turn now to the measurement of images intended for *viewing*, as distinct from negatives of which the visual appearance is not ordinarily important.

34. Images as seen by reflected light (i.e., paper prints), may be described in terms of one of two numbers. One is the *reflectance*, which is similar to *transmittance* for negatives. Reflectance is a fraction, like transmittance; it is the fractional part of the light received by the sample that the sample reflects back. If the sample receives 100 units of light and reflects 10 units, its reflectance is __________. — 0.10
35. If a sample of a print receives 200 units of light and reflects 2 units, its reflectance is __________. — 0.01
36. A very nearly white paper base may reflect 190 units of light if it receives 200 units. Its reflectance is __________. — 0.95
37. The darkest possible area of a print reflects only very little light, perhaps only 1 unit if it receives 200 units. The reflectance of such an area would be __________. — 0.005
38. A numerical concept like opacity does not exist for reflecting images. We do, however, define reflection *density* just as for transmitting images. Reflection density is the log of the reciprocal of the reflectance. Thus, if the reflectance of a sample is 0.10, we first find the reciprocal, that is, 1 divided by 0.10, or __________. — 10
39. We then find the log of the reciprocal (10) to be __________. — 1.00
40. If the reflectance is 0.05, we follow the same procedure: find the reciprocal, which is __________ and then the log of the reciprocal, which is __________. — 20, 1.30
41. If the reflectance of paper stock is 0.95, its density is only __________. — 0.02

42. The maximum possible reflection density would be for an area of least possible reflectance (that is, about 0.005). The maximum density for a paper print is therefore about __________ . — 2.30

43. We ordinarily prefer density (a logarithm) as a measure of print images because density is approximately related to the *appearance* of the image. If we present to an observer a large number of different grays—light, medium, and dark—and ask the observer to select a set of visually equally-spaced grays, we find that the densities of the selected set are also about equally spaced. Thus equal differences in visual appearance are suggested by equal differences in __________ . — density

44. If two areas of a print have densities of 0.30 and 0.60, a third area as different from the second as the second is from the first would have a density of about __________ . — 0.90

45. The corresponding print reflectances for the three areas would in order be __________ , __________ *and* __________ . — 0.50, 0.25, 0.125

46. Note that although the *densities* of the three areas are equally spaced, the reflectances are unequally spaced since the difference in the first two reflectances is 0.25, but the difference in the last two is only __________ . — 0.125

47. Thus if we want to describe visual images in terms of numbers, it is logical to use logarithmic values, that is, __________ . — density

48. In the zone system, there are said to be 10 zones in a full-range print; zones are about equally spaced in the visual sense. If a normal print has a range in density of about 2.0, each zone differs from the next by a density value of about __________ . — 0.20

49. The human eye is exceedingly sensitive under good viewing conditions: you can detect a difference between two grays when the density difference is only about 0.01. In a full-range print, with a total density difference of about 2.0, the number of potentially different grays is thus about __________ . — 200

SELF-TEST ON INTRODUCTION TO THE CHARACTERISTIC CURVE

Check your understanding of this section by answering the following questions. The correct answers follow.

1. If a sample receives 50 units of light and transmits 5 units, its transmittance is __________ .
2. If the transmittance of a negative sample is 0.5, its opacity is __________ .
3. If the opacity of a sample negative is 4, its density is __________ .
4. If the density of a sample is 1.0, its opacity is __________ and its transmittance is __________.
5. One reason for preferring density (rather than transmittance or opacity) as a measure of the photographic image is that density is directly proportional to the amount of __________ in the image.
6. The reflectance of an area of a print is 0.25. Its reflection density is __________ .
7. We use density as a measure of an image intended for viewing because density is approximately related to the __________ of the image.

ANSWERS TO SELF-TEST ON INTRODUCTION TO THE CHARACTERISTIC CURVE

In the parentheses that follow each answer you find the number of the item in the preceding section that relates to the question and its answer.

1. 0.1 (items 1–4).
2. 2 (items 8–15).
3. 0.60 (items 19–22).
4. 10, 0.1 (items 23–25).
5. Silver (items 26–28).
6. 0.60 (items 38–41).
7. Appearance (items 43–47).

In the first part of this chapter we dealt with the concepts associated with the measurement of the photographic image. We turn now to the basic concepts related to the *exposure* of the image and thus to the numbers appearing on the horizontal axis of the *D*–log *H* curves at the beginning of this chapter. You saw there that exposure means the *quantity* of *light* that is received by an area of a photosensitive material (film in the camera or photographic paper in the printer).

WHAT YOU WILL LEARN

If you work carefully with the following programmed material, you will:

1. Understand the meaning of the term "light";
2. Know how to find the value of the exposure;
3. Know how exposures in the camera are related to the tones of the subject;
4. Know how to find the contrast of a subject;
5. Know why exposure is expressed in logarithms;
6. Know when it is necessary to use other than normal development times;
7. How exposures in the camera are changed when the camera aperture and shutter settings are changed;
8. The meaning of "stops" in exposure.

DIRECTIONS. In order for you to understand this material, it is important for you to follow the instructions on page 35 at the beginning of this chapter.

50. The term "light" means that kind of radiation capable of affecting the normal human eye. Other kinds of radiation that do not permit us to see are *not* light. Ultraviolet radiation is commonly called "black" light. It is invisible, and therefore by our definition ________ really light.
    is, is not

    is not

51. Infrared radiation can affect some kinds of photographic film, but it is invisible. Infrared is therefore not ____________ .

    light

52. Some lasers produce radiation that we cannot see. Such lasers __________ (give, do not give) light. — do not give

53. Some photoelectric meters used by photographers respond to ultraviolet and infrared and therefore could possibly give a reading in complete darkness. Such meters are __________ (truly, not truly) *light* meters. — not truly

54. Some stars give out a large quantity of radiation that cannot be seen, although it can be detected by other means, including photography. Such stars give out practically no __________ . — light

55. To measure light, a meter must "see" as does the eye. Only a specially designed photoelectric meter does so. Furthermore, exposure means the amount of light received at the *film;* therefore any meter reading of the light falling on (or coming from) the subject outside the camera is __________ (measuring, not measuring) exposure as we have defined it. — not measuring

56. Therefore the photographer's "exposure" meter as customarily used is not really measuring exposure in this sense. To know the amount of exposure as here defined, the measurement would need to be made __________ (inside, outside) the camera. — inside

57. The amount of exposure is controlled by two factors: (a) the *time* during which the light hits the film, usually controlled by the shutter and (b) the strength of the light on the film. When a leaf (between-the-lens) shutter is used, the time is constant all over the film. The *strength* of the light, however, will differ in various positions on the film; it is controlled in part by the *subject,* each different part sending a different quantity of light to the camera lens and thus to the film. Dark shadows, for example, send only a small quantity of light to the camera, and the exposure for the shadows will thus be __________ . — small, weak (etc.)

58. Bright highlights, however, send a large amount of light to the camera, and therefore produce __________ exposures. — large, great (etc.)

59. Subject areas (tones) of intermediate brightness (midtones) send moderate quantities of light to the camera, and thus produce exposures of a moderate level. Every ordinary subject (like a person or a landscape) contains many different tones—shadows, midtones, and highlights—and thus for such subjects there will exist at the film __________ (many, few) different exposures. — many

60. Usually, therefore, when a single picture is made of a typical subject, there will be an uncountable number of different exposures at the film. In some cases, however, there will be only a small number of different exposures. In microfilming, for example, a document like this page of text is photographed. The page has only two different tones—the paper and the characters. In this case, a photograph would involve only __________ different exposures. — two

61. In microfilming this page, it would be uniformly lighted. The paper would reflect much light to the camera, and thus produce a __________ exposure. — large, great (etc.)

62. For the same situation, however, the dark ink of the characters would reflect only a little light, and thus would produce only a __________ exposure. — small (etc.)

63. If it were photographed, a uniformly lighted surface like a flat painted wall would

produce only one exposure. A poster made up of black and gray ink on white paper would produce __________ different exposures. — three

64. For the poster, the largest exposure would be produced by the __________ part of the subject. — white
65. The smallest exposure would come from the __________ part of the poster. — black
66. In item 57 above you saw that exposure is fixed by the time and by the strength of light at the film. Exposure is defined as the product of these two factors: $H = E \times t$, H standing for the *exposure*, E for the light level (the technical word is *illuminance*), and t for the time. If at one part of the film the illuminance is 25 units and the time 1/10 unit, the exposure is __________ units. — 2.5
67. If at another point on the film the illuminance is 100 units and the time is the same as before, the exposure there is __________ units. — 10
68. Thus we *compute* (we almost never measure) the exposure from the illuminance and the time. The illuminance at the film is different for every different part of the subject. It can be measured, but only with great difficulty. What *can* be, and is often, measured is the light sent from the different tones of the subject toward the camera. The common term for this type of measurement is "reflected-light" (or "reflectance") meter reading, although the light need not be reflected; it could come directly from a light source in the field of the camera. Such a measurement is *not* of exposure nor of illuminance *on* the film, but a value *related to* the light on the film in this way: if two parts of the subject give "reflected-light" measurements in a *ratio* of 10 to 1, the two illuminances on the film will have nearly the *same* ratio. If the time of exposure is constant, the exposures will also have a ratio of __________ . — 10 to 1
69. If you find the extreme tones of a subject, that is, the brightest highlight and the darkest shadow and measure these with a "reflected-light" meter, you can find from the readings the ratio of illuminances on the film and thus the ratio of exposures. If the darkest shadow gives a reading that implies 4 units of light and the highlight reading represents 1000 units of light, the ratio of exposures on the film would be __________ . — 250 to 1
70. In making a portrait, perhaps readings on the two sides of the subject's face would be interpreted as meaning 30 and 12 units of light; the ratio of exposures on the film for these two tones would be __________ . — 2½ to 1
71. Such a ratio is a good measure of the *subject contrast*. If "reflected-light" measurements imply 20 and 500 units of light for the extreme subject tones, the subject contrast would be __________ . — 25 to 1
72. For the same situation, the extreme ratio of exposures on the film would be __________ . — 25 to 1
73. In the section above on density, you saw that we prefer a logarithmic measure of the print because equal log differences imply nearly equal visual differences. For the same reason, we usually prefer a logarithmic measure of *exposure*, because equal differences in exposure imply nearly equal differences in the tones of the subject that produce those exposures. For example, in the Adams-White zone system, each zone is equivalent to a shade of gray. In the subject, each zone (identified by a Roman numeral from I to IX) differs from the next by 0.3 in logs. A subject that extends from zones II to VII has five such intervals, for a total log difference in the subject of __________ . — 1.5

74. Most modern light meters have logarithmic scales. The reason for such scales is that equal log intervals represent about equal __________ differences in the subject. — visual

75. On this scale, the number values are arbitrary, and are usually 1, 2, 3, . . ., 19, 20, 21. Each scale interval means a log difference of 0.3, just like the intervals in the zone system. When such a meter is pointed at different parts of the subject, it measures different tones (or zones) of the subject. If three readings made at different parts of the subject give values of 5, 7, and 9 (i.e., equally spaced values), we can suppose that the three areas of the subject appear ______ (equally, unequally) different. — equally

76. If a different subject gave readings of 5, 7, and 11, we would infer that these three tones would appear ______________ (equally, unequally) spaced. — unequally

77. When we use a logarithmic measure of exposure (as on the horizontal axis of the characteristic curve), we can mentally transfer such subject readings directly to that axis. For every different tone (zone) of the subject there will be a different exposure. The log differences in the subject tones are (if the camera optical system is good) transferred as log-exposure differences on the film. Thus if two tones of the subject measure 7 and 11 on this type of meter, the difference in these values is 4. Each interval is 0.3, and thus the total difference is __________ . — 1.2

78. These two tones of the subject would produce a log-exposure interval equal to __________ . — 1.2

79. If you were to make a picture of this subject, the two tones would be represented by two tick marks on the horizontal axis of the characteristic curve a log distance of __________ apart. — 1.2

80. If in another subject there were two zones measuring 15 and 21, the difference in these values is 6. Since each scale interval is worth 0.3 in logs, the total log difference is __________ . — 1.8

81. If you were to photograph this subject, the log-exposure interval on the horizontal axis of the characteristic curve would be __________ . — 1.8

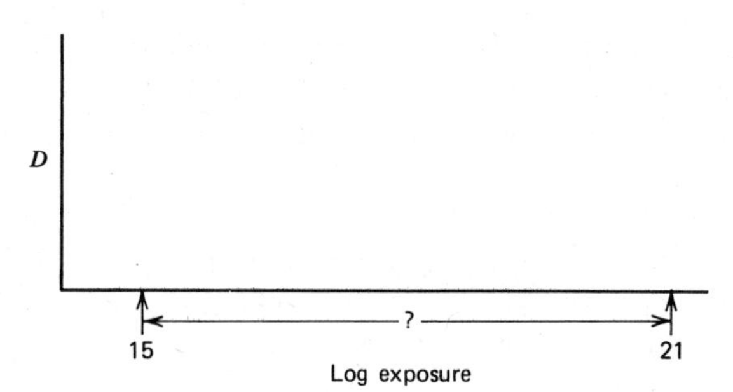

82. A still different subject might give readings of 11 and 21 for a total log difference of __________ . 3.0

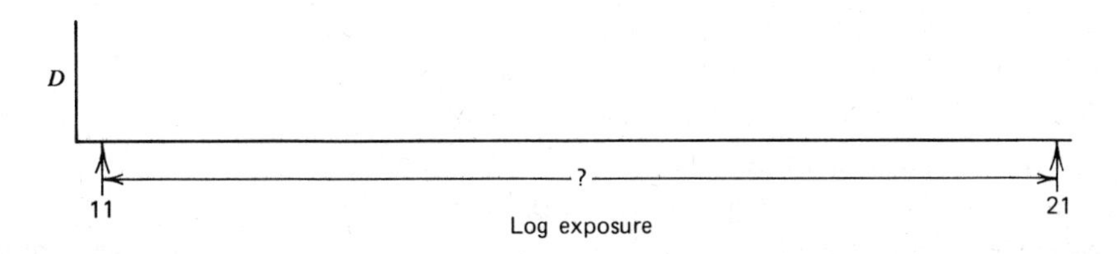

83. It is important to know the total range of the subject, because knowledge of this number lets you know about the likelihood of success in reproducing the subject. A *typical outdoor* subject in good lighting has a range of about 7 scale divisions from darkest to lightest tone. This corresponds to a log-exposure interval of __________ . 2.1

84. A foggy-day scene outdoors may produce a total log range of only about 1.2, about __________ meter scale divisions. 4

85. A winter snow scene, in bright sunlight with heavy shadows, may produce a log-exposure interval as large as 3.9, or __________ scale divisions. 13

86. A typical outdoor subject will have a range of about 7 scale divisions; a "flat" subject considerably fewer, say 5 or fewer; a "contrasty" subject considerably more, say more than 8.

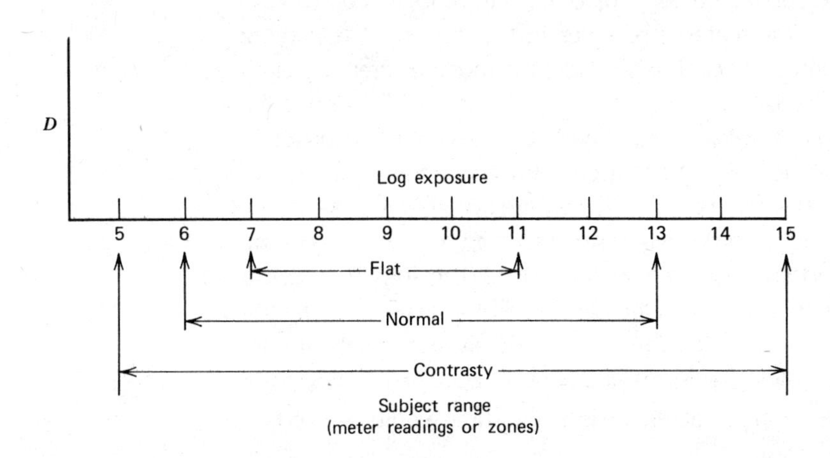

Normal methods will produce excellent negatives of typical subjects; modified methods will be needed for unusually flat or contrasty subjects. If the darkest shadow measures 4 and the lightest highlight 10, you would expect to use ______________ procedures.
normal, modified

normal

87. A commonly used procedure is supplementary lighting (such as fill-in flash) when the subject is too contrasty. Adding light to the shadows makes them closer to the highlights. If the deepest shadow reads 8 and the lightest highlight reads 15, fill-in flash ______________ be needed.
would, would not

would not

88. On the other hand, for a subject giving extreme readings of 5 and 15, fill-in flash ______________ be helpful.
would, would not

would

89. A different modification of the process, commonly used in the zone system, is to change the development time (from normal) when the subject contrast is found to be unusual. Development time is increased for a flat subject (one with a range of five or fewer zones) and decreased for a contrasty subject (one with a range of 8 or more). If the subject gave readings from 4 to 12, you would, using this method, develop the negative for a time that is ______________
normal, more than normal, less than normal.

less than normal

90. If the subject gave extreme readings of 8 and 12, using this method you would develop the negative for a time that is ______________.
normal, more than normal, less than normal

more than normal

91. If the subject gave extreme readings of 7 and 19, using this method you would develop the negative for a time that is ______________.
normal, more than normal, less than normal

less than normal

92. As you saw in items 77–86 above, the *range* of *tones* of the subject controls the *range* of *log exposures* that the material receives in the camera. The camera settings [*f*-number (aperture) and time] fix the actual exposures themselves as specific log values on the horizontal axis of the *D*–log *H* curve. Thus once the shutter is operated the log *H* values are determined. If another film is exposed in the camera at different values of the *f*-number and time, the entire set of exposures (from the subject) is moved *sideways* on the horizontal axis—to the *right* if the time is *more* or the aperture *greater;* to the *left* if the time is *less* or the aperture *smaller.* One stop change is equivalent to 0.30 in logs (because a change of one stop represents a change in the light level by a factor of 2, and the log of 2 is 0.30.) Assume a subject with a range of 6 scale divisions on the meter; it would produce a range of log exposures of ______________ .

1.8

93. For a given camera setting, say 1/100 at *f*/8, the log exposures would be fixed, as shown in the following diagram:

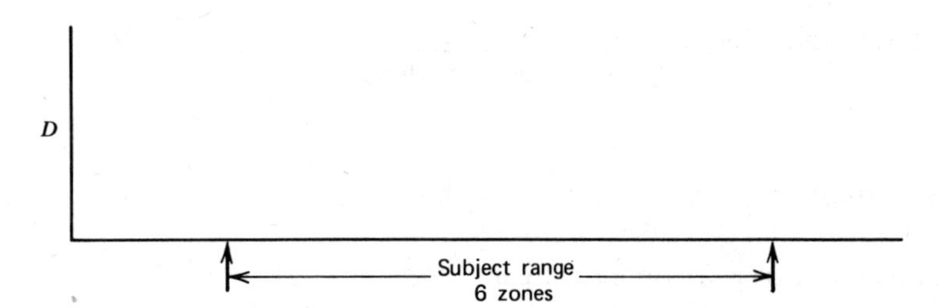

If you were to make a second negative, with an aperture one stop greater (say 1/100, at *f*/5.6), the exposures produced by the subject would be moved to the __________, a distance of __________ in logs. — right, 0.30

94. Thus comparison of the second to the first exposing situation would be as shown:

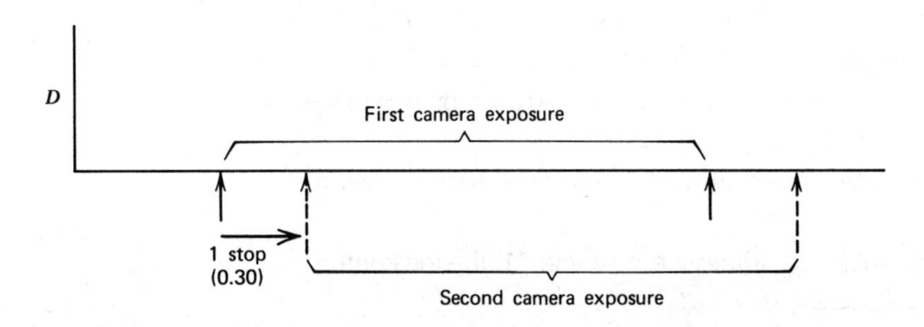

"Opening up" the lens aperture by two stops instead of one would have moved the set of exposures just twice as far as before, specifically, a distance in logs of __________. — 0.60

95. A half stop is just half as much (in logs) as a full stop, that is, a half stop is equal to a sideways shift of __________. — 0.15

96. Increasing the *time* of exposure by a factor of 2 has the same effect on the exposure set as opening up the lens by a stop. If the shutter setting is changed from ½ to 1 second, a correctly operating shutter would shift the subject exposures to the __________ a distance of __________ in logs. — right, 0.30

97. Opening the aperture (or increasing the time of exposure) moves the whole set of subject exposures to the __________. — right

98. Closing down the aperture (or decreasing the time of exposure) moves the whole set of subject exposures to the __________. — left

99. Every stop change moves the subject exposures a distance of __________ in logs. — 0.30

SELF-TEST ON INTRODUCTION TO THE CHARACTERISTIC CURVE (2)

Check your understanding of the preceding section by answering the following questions. The correct answers follow the questions.

1. Light is radiation that affects the normal __________.

2. Photoelectric meters usually respond to radiation ____________ (in the same way as, differently from) the human eye.
3. Strictly speaking, exposure on a film must be found ____________ (inside, outside) the camera.
4. The two factors that govern the amount of exposure on the film are ____________ and ____________.
5. When a single negative is made of a typical subject, there is/are on the film ____________ (one, several, many) exposure(s).
6. To compute exposure we use the formula: $H =$ ____________ $\times$ ____________.
7. If the ratio of reflected-light measurements from a subject is 20 to 1, the ratio of exposures for that subject on the film will be nearly ____________ to 1.
8. If the extreme tones of a subject send, respectively, 10 and 200 units of light to the camera, the subject contrast would be ____________ to 1.
9. On most modern photoelectric meters, one scale division (as from 10 to 11) means a log difference of ____________.
10. On such a meter, if the extreme subject readings were 14 and 21, the log contrast of the subject would be ____________.
11. For an outdoor subject, readings such as in 9 would represent a ____________ (typical, flat, contrasty) subject.
12. For a negative made of a subject with a log contrast of 3.0, for ease in printing development should be ____________ than normal.
13. For a subject with a log contrast of 1.5, for ease in printing development should be ____________ than normal.
14. Two negatives are exposed of the same subject, the second with one stop more exposure in the camera than the first. In the second case, the exposures representing the subject have been moved on the log exposure axis to the ____________.
15. In the same situation as in 14, the distance moved is in logs ____________.

ANSWERS TO SELF-TEST ON INTRODUCTION TO THE CHARACTERISTIC CURVE (2)

In the parentheses that follow each answer you find the items in the preceding section that relate to the question and its answer.

1. Human visual system, or human eye (items 50–54).
2. Differently from (items 53, 55).
3. Inside (items 55–56).
4. Illuminance (light strength) and time (items 57, 66).
5. Many (items 57–59).
6. $H = E \times t$ (item 66).

7. 20 to 1 (items 68–70).
8. 20 to 1 (items 71–72).
9. 0.3 (items 75, 80–85).
10. 2.1 (items 80–86).
11. Typical (item 86).
12. Less (items 89–91).
13. More (items 89–91).
14. Right (items 92–97).
15. 0.3 (items 93, 96, 99).

CHAPTER 3

How to Read a Logarithmic Scale

In Chapter 2 you were introduced to the characteristic curve, a plot resulting from a test of a photographic material. You saw that logarithmic scales are ordinarily used in such plots. Elsewhere in photography log scales are also commonly used. Examples appear in the introduction to Appendix Section A1.

Logarithmic scales, as you have seen, are used because of their relation to the process of vision, and because such scales properly represent data in which number ratios are important. Being able to read values on a log scale is often necessary. The following assumes that you are familiar with Appendix Sections A3 and A4.

WHAT YOU WILL LEARN

If you work carefully through this chapter, you will be able to identify correctly any value on a scale marked in logarithms.

DIRECTIONS. To obtain the skills needed to work easily with a log scale, you need to follow this procedure faithfully: cover about 2 inches of the right hand margin of each of the following pages in turn with an opaque sheet of paper. Read carefully the first numbered statement. *Write* the answer that you believe correctly completes the statement. Move the cover sheet down to reveal the correct answer in the margin. Continue in this way until you have finished the section. Work at a rate that is comfortable for you, and no longer than you find pleasant. Begin when you wish.

1. Note carefully the following logarithmic scale:

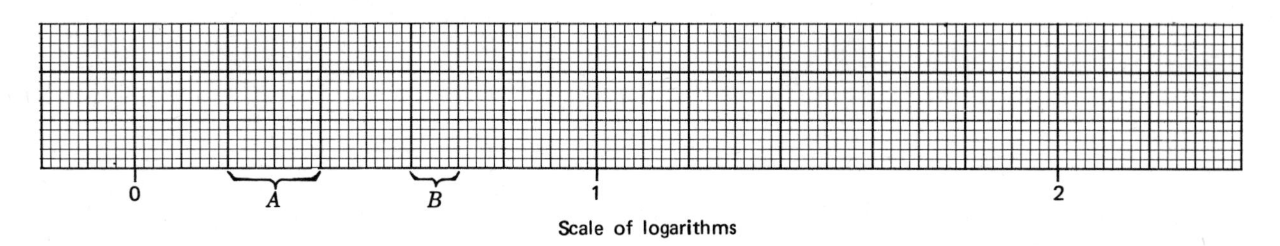

The whole-number values 0, 1, and 2 represent that part of a logarithm called the ____________ . — characteristic

2. Between 0 and 1 there are five intervals each of width equal to that of bracket *A*. Each such interval has the value ____________ . — 0.2
3. The interval represented by bracket *B* has the value ____________ . — 0.1
4. The smallest subdivision on this scale has the value ____________ . — 0.02
5. The smallest subdivision has the value 0.02, because five such subdivisions make up the interval of 0.1 represented by the bracket *B*. By reading off the decimal values between the characteristics we find the part of the logarithm called the ____________ . — mantissa
6. Consider the following similar scale:

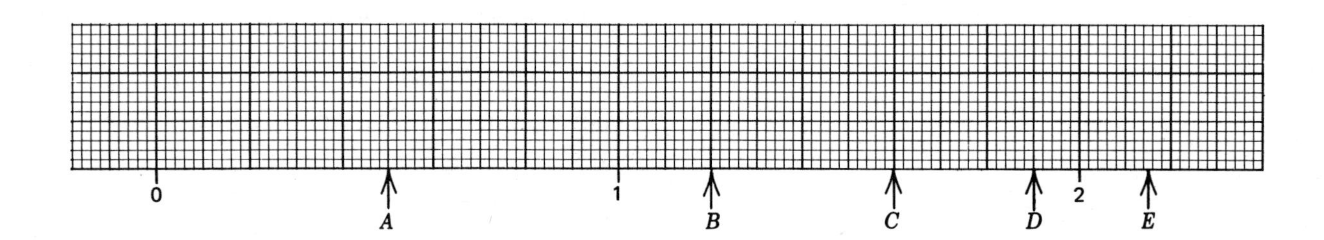

In the range of values shown here, values are read just as from a ruler. Arrow *A* is midway between 0 and 1 and thus represents the value ____________ . — 0.5

7. On the same scale, arrow *B* has the value ____________ . — 1.2
8. Arrow *C* has the value ____________ , *D* the value ____________ , and *E* the value ____________ . — 1.6, 1.9, 2.15
9. Note the convenience of this scale, where each inch has the value 0.4, each ½ inch the value 0.2, and each of the smallest subdivisions the value ____________ .Using such a scale, we can estimate values to about 0.01. — 0.02
10. Note especially that between the points marked 1 and 2 the whole-number part of the logarithm remains the same; it is the characteristic ____________ . — 1
11. Exactly the same pattern applies to the part of the scale to the left of the zero (0) position. Consider the following scale:

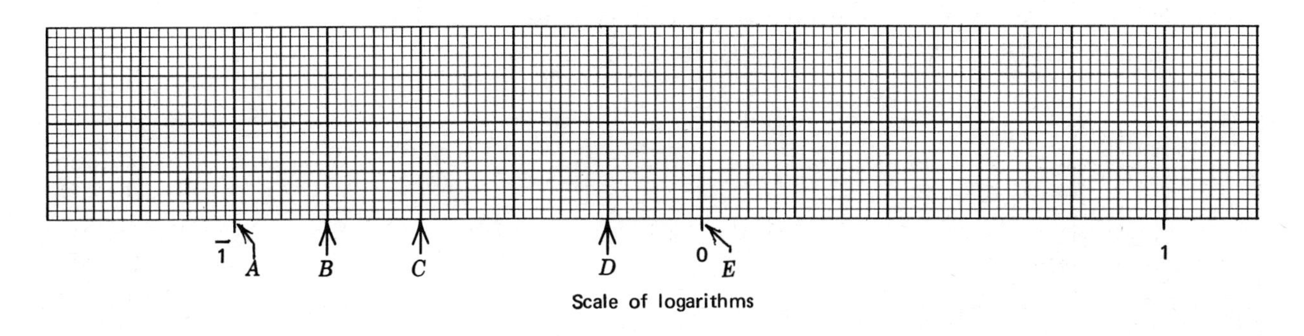

Scale of logarithms

In the region of this scale from $\bar{1}$ to 0, the characteristic of the log is the same as at point *A*, that is, ___________ . — $\bar{1}$

12. At point *B* we retain the same characteristic ($\bar{1}$), add 0.2 for the new mantissa, and write ___________ . — $\bar{1}.2$
13. Point *C* has the same characteristic as before, and 0.2 more for the mantissa, so the reading is ___________ . — $\bar{1}.4$
14. Point *D* has the value ___________ . — $\bar{1}.8$
15. For point *E* we add 0.2 to the mantissa that we had for point *D* (0.8) and get ___________ . — 1.0
16. This positive 1, combined with the characteristic of $\bar{1}$ makes the net value of the entire logarithm ___________ . — 0
17. The same sequence applies to the region between $\bar{2}$ and $\bar{1}$ as in the following scale:

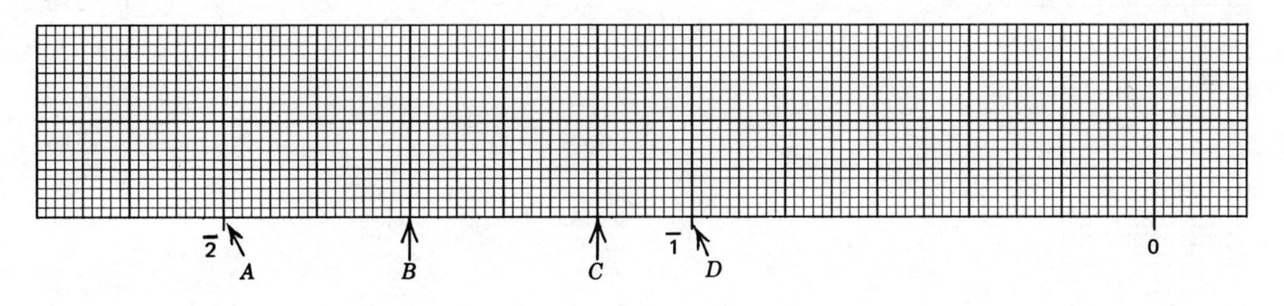

Over the range from $\bar{2}$ to $\bar{1}$, the characteristic is the same as at point *A*, that is, ___________ . — $\bar{2}$

18. Point *B* has the value ___________ and point *C* the value ___________ . — $\bar{2}.4$, $\bar{2}.8$

Over any given interval of whole numbers (characteristics) only the mantissa of the log increases; the characteristic remains the same.

SELF-TEST ON HOW TO READ A LOG SCALE

Cover the right-hand margin of the page with an opaque piece of paper. Identify the values associated with each of the lettered arrows. *Write* your answer on the covering

sheet of paper; then move the paper down until the correct answer is revealed.
NOTE THE BARS OVER SOME OF THE CHARACTERISTICS.

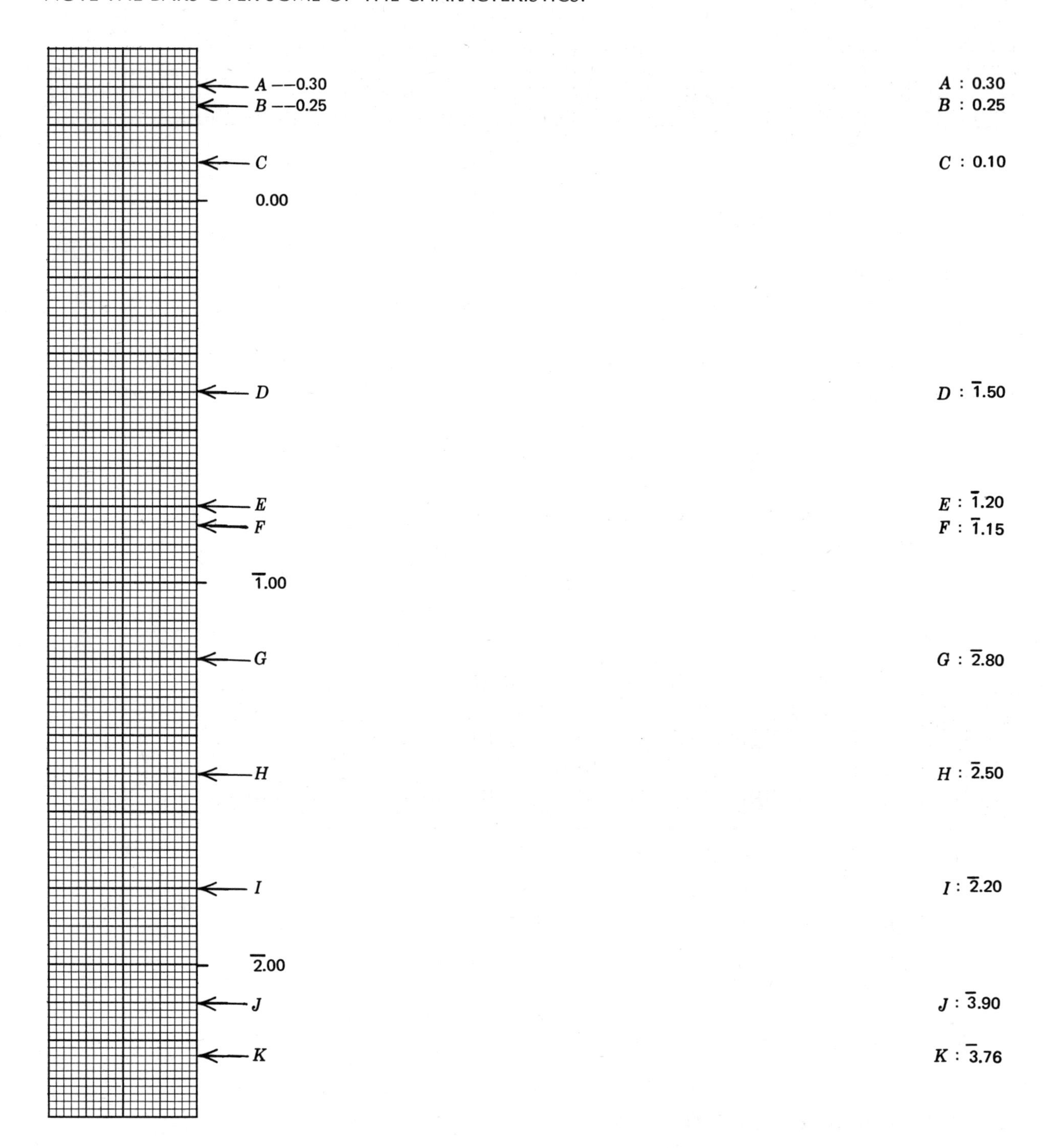

CHAPTER 4
The Evaluation of the Negative *D*–log *H* Curve

When a sample of photosensitive material has been appropriately exposed, processed and measured, the plotted *D*–log *H* curve describes the relationship between the response of the material and the exposure and the processing it received. A generalized *D*–log *H* curve is shown on page 54.

On the vertical axis is shown the density—a logarithmic measure of the image that was produced. On the horizontal axis is displayed the exposure, also in logarithms. Log scales are conventionally used for both axes for the reasons explained in Chapter. 2.

The data needed to plot such a curve come most directly from a test in which a sensitometer (a precise exposing device) has been used to give known exposures on the sample film. The concepts to be developed in this chapter, however, also apply to exposures made in a camera.

From such a *D*–log *H* curve we can find:

1. The range of log exposure values that the material can accept. This range is closely related to the extreme range of subject tones that can be recorded. For a subject of smaller range, it also indicates the allowable tolerance ("latitude") in camera settings.
2. The range of image densities that the material can produce. This range is an indication of the greatest contrast that can be found in the image.
3. The "speed" of the material, and from this value the necessary camera settings for a given light level.

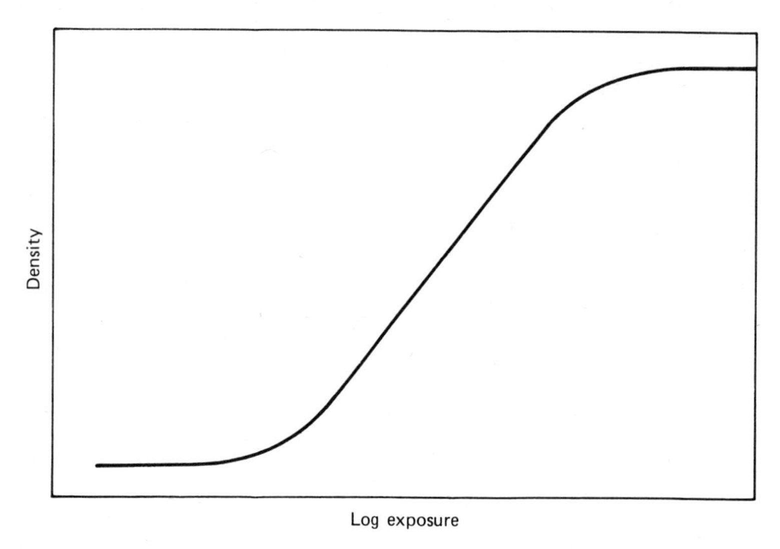

4. The "contrast" of the material, which expresses the relationship between the image and the subject as recorded by the material.

These four properties of the photosensitive material are greatly influenced by the way the sample was exposed, processed, and measured. They are by no means basic to, or fixed by, the material itself. Nevertheless, these properties are important to the photographer because knowledge of them is fundamental to the intelligent choice and use of the photographic emulsion.

To understand the concepts involved in the *D*–log curve, you need to appreciate the importance of the *shape* of the curve. Curve shape is indicated by the *steepness* of the curve. Its steepness expresses how the image (measured by the densities on the vertical axis) relates to the subject exposures (found on the horizontal axis).

WHAT YOU WILL LEARN

By working carefully with the following programmed material, you will:

1. Understand that image quality is related to the steepness of the *D*–log *H* curve;
2. Be able to measure the steepness of any part of the curve;
3. Know how to measure the contrast of the image.

DIRECTIONS. Cover about 2 inches of the right-hand margin of each of the following pages in turn with an opaque sheet of paper. Read carefully the first numbered statement. *Write* the word or phrase that you believe correctly completes the statement. Move the cover sheet down to reveal the correct answer in the margin. Continue in this way until you have finished the section. Work at a rate that is comfortable for you, and no longer than you find pleasant. Begin when you wish.

What is important about the image is the *difference* in the image densities. The density difference is a measure of image *contrast* and of image *detail*. To be meaningful, the image (density) difference must be compared with the difference in the subject exposures that produced the image.

1. Consider the following steep curve.

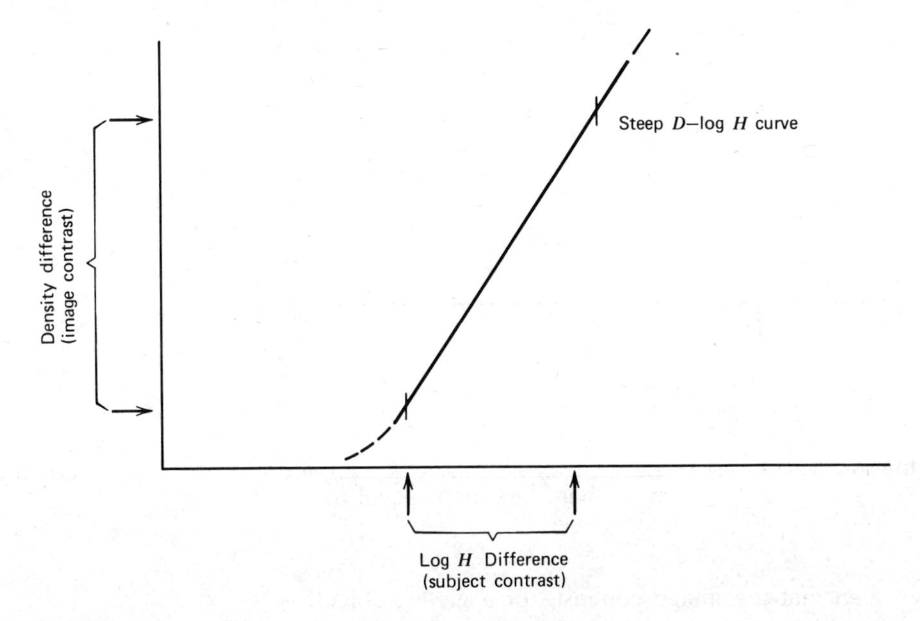

In comparison with the subject log-exposure difference, the image difference is __________ small, large. — large

2. Now consider the following flat curve.

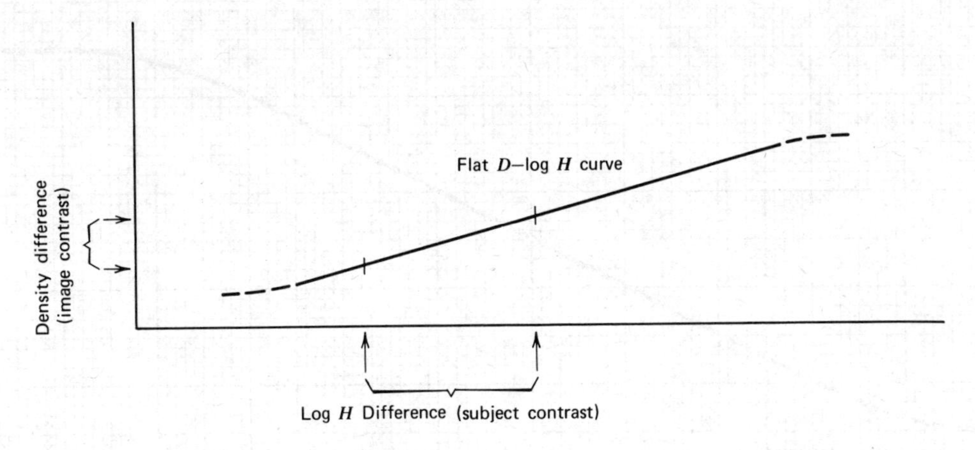

For the *same* subject log exposure difference as before, the image difference is now __________. small, large — small

3. Now consider the following curve. This is a special case; the central part of the curve makes an angle of 45° with the horizontal axis.

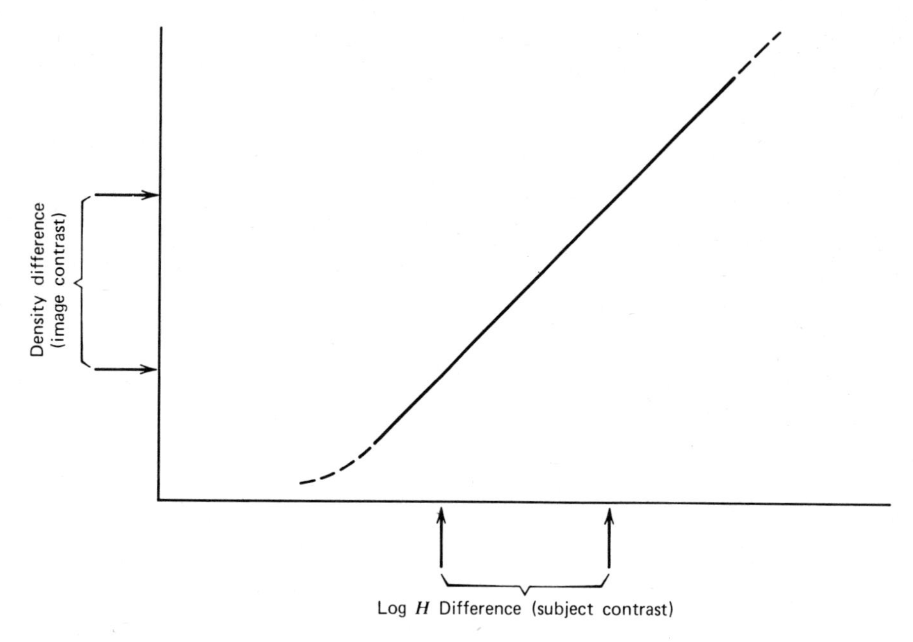

For this special case, the image contrast is ______________ the subject contrast.
(more than, less than, equal to)

equal to

In the preceding, you have seen that the image contrast, for a given subject, is determined by the steepness of the D–log H curve on which the subject log exposures fall. The numerical measure of steepness is called the *slope*. In items 4–21 that follow, refer to the D–log H curve below.

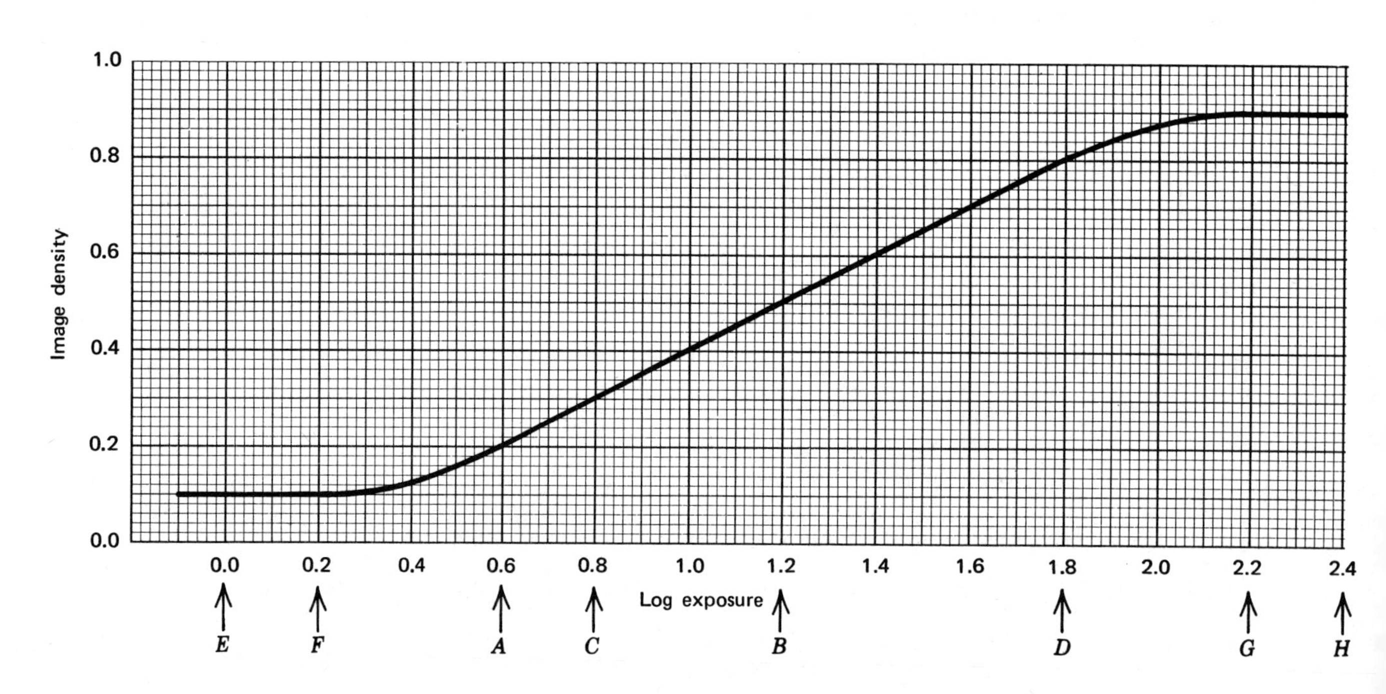

4. Consider the two points *A* and *B*, representing two subject exposures. On the vertical axis, read off the density for point *A*. The density is ______ . — 0.20
5. Read off the density corresponding to point *B*. It is ______ . — 0.50
6. Find the difference in these two densities. It is ______ . — 0.30
7. Now read off the log exposure values for points *A* and *B*. The values are ______ and ______ . — 0.60, 1.20
8. Find the difference in the log-exposure values. It is ______ . — 0.60
9. Finally, divide the density difference (answer to item 6) by the log-exposure difference (answer to item 8). The result is ______ . — 0.50
10. You have found the *slope* of the *D*–log *H* curve between points *A* and *B*. The result, 0.50, means that the image contrast for these two subject exposures will be *half* as great as the subject contrast. Now consider the points *C* and *D*. Find the density difference for these points. It is ______ . — 0.50
11. Find the log *H* difference for the same points. It is ______ . — 1.00
12. Now find the slope, the ratio of your answers to items 10 and 11. The slope is ______ . — 0.50
13. These two examples should convince you that if you choose *any* two points on the *straight line* part of the *D*–log *H* curve, you will always obtain the same slope, in this case a value of ______ . — 0.50

Because the straight line has a constant slope, the image contrast will be the same for a subject exposed on a given straight line, no matter where the subject exposures fall.

14. Return to the *D*–log *H* curve (page 56) and consider points *E* and *F*. The log *H* difference for these points is ______. — 0.20
15. For the same two points, the density difference is for all practical purposes ______ . — zero
16. The slope, the ratio of the density difference to the log *H* difference is 0.00/0.20, or ______ . — zero
17. If, over some log *H* interval, the curve does not rise at all, the slope of the curve there is always ______ . — zero
18. That the slope is zero means that there is *no* difference in the image tones, even though there was a real difference in the subject tones. For subjects exposed on the horizontal part of the *D*–log *H* curve, the image will be completely lacking in ______ . — contrast, detail, difference in tones (etc.)
19. For the same curve, consider points *G* and *H*. The slope of the curve between these points is ______ . — zero
20. If any part of the subject is exposed on this region of the *D*–log *H* curve, the image of that part of the subject will have ______ contrast (detail). (no, a little, much) — no
21. The steepest part of the *D*–log *H* curve is the straight line. Since the slope is a maximum in the straight line, subject exposures falling on that part of the *D*–log *H* curve will be reproduced in the image with the greatest amount of ______. — detail, contrast (etc.)

For items 22–28 refer to the following D–log H curve.

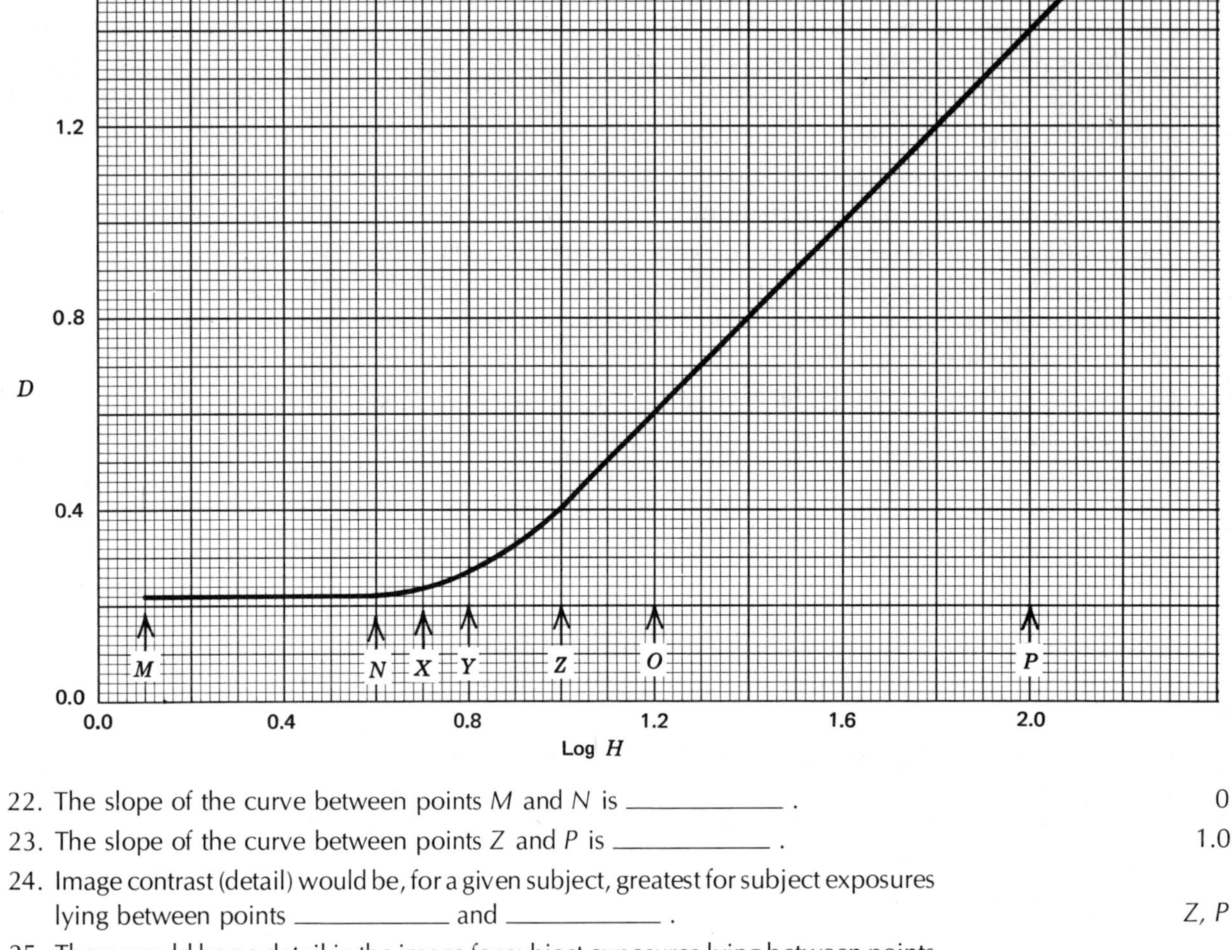

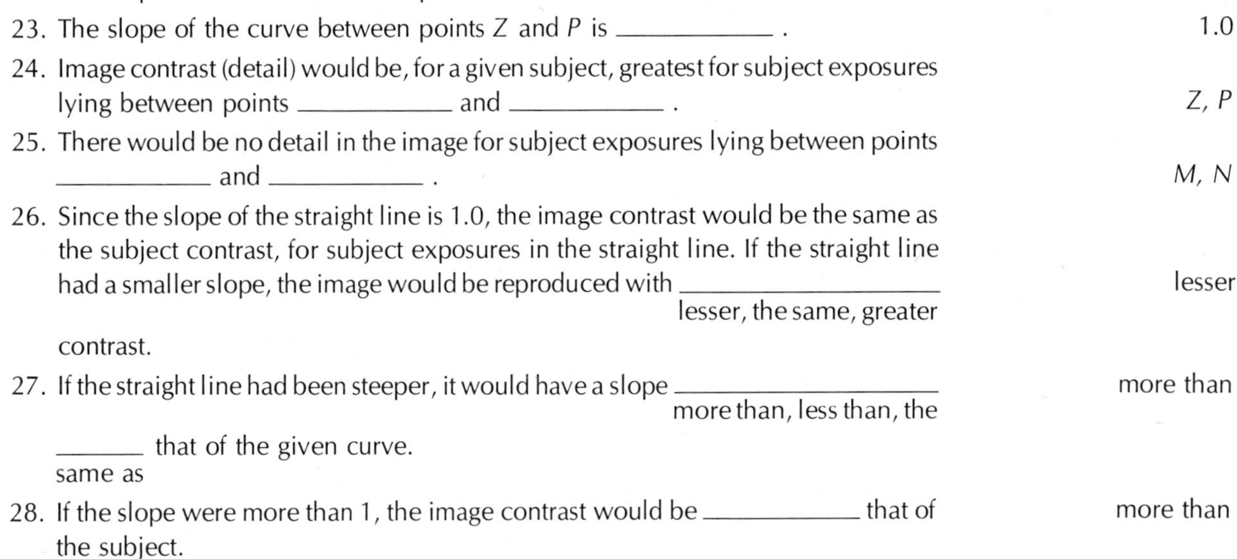

22. The slope of the curve between points M and N is ________ . — 0
23. The slope of the curve between points Z and P is ________ . — 1.0
24. Image contrast (detail) would be, for a given subject, greatest for subject exposures lying between points ________ and ________ . — Z, P
25. There would be no detail in the image for subject exposures lying between points ________ and ________ . — M, N
26. Since the slope of the straight line is 1.0, the image contrast would be the same as the subject contrast, for subject exposures in the straight line. If the straight line had a smaller slope, the image would be reproduced with ________ (lesser, the same, greater) contrast. — lesser
27. If the straight line had been steeper, it would have a slope ________ (more than, less than, the same as) that of the given curve. — more than
28. If the slope were more than 1, the image contrast would be ________ that of the subject. — more than

In the D–log H curve in the preceding plot, the region between points N and Z is not straight. It is called the "toe" of the curve. In this part of the curve, the steepness, and therefore the slope, changes gradually, increasing from left to right. Because of the gradual change in the slope, we cannot choose *any* two points here. If, for example, you select X as one of the points, you will get a different slope if you choose Y or Z as the other point.

We solve this difficulty by finding the slope *at* a single point, as in the following items (refer to the following curve). We desire the slope at point Q. We have drawn a

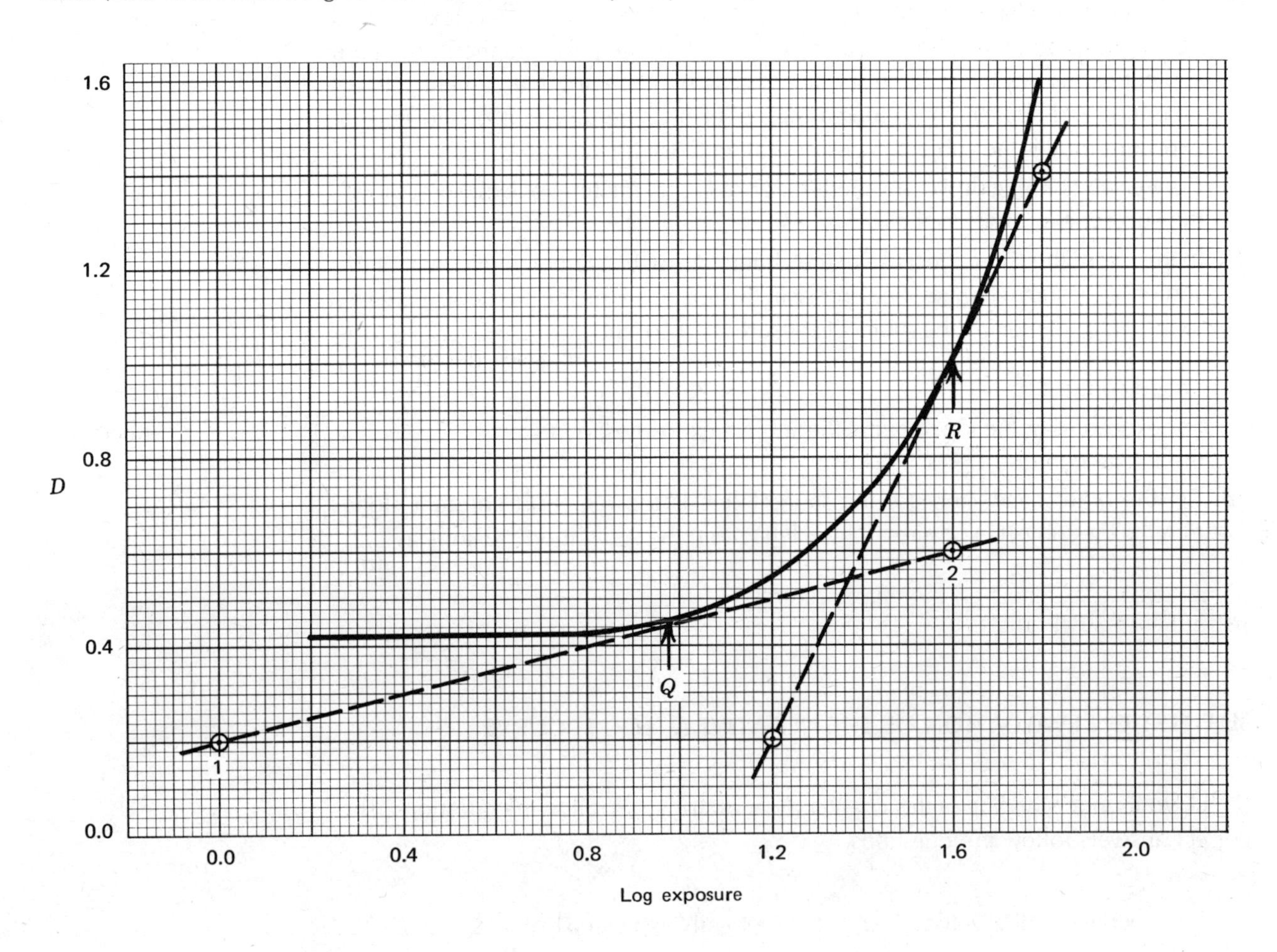

straight dotted line *tangent* (just touching) the curve at that point.

29. Use points 1 and 2 (circled on the tangent line) to find the slope of the straight-line tangent: the density difference is ________ ; the log H difference is ________ ; the slope is ________ .

0.40
1.60 0.25

30. You found that the slope of the tangent at point Q is 0.25. Since the tangent and the D–log H curve coincide at point Q, you have found that slope of the curve at that *same* point. That the slope is approximately one-fourth means that a subject tone exposed at this spot on the D–log H curve would have a contrast in the image about one-fourth that in the ________ .

subject

31. Refer to point R on the same curve (preceding diagram). By the same procedure as before, find the slope at this new point. The slope of the curve at point R is about ________ .

2.0

32. That the slope is 2.0 at point R means that if a subject exposure lies at this point, the image contrast is ____________________ the subject contrast.
more than, the same as, less than

more than

33. Starting from the lefthand part of this curve, and moving to the right, you see that the slope continually ____________ .
 increases, stays the same, decreases

 increases

34. For this *D*–log *H* curve, farther yet to the right past point *R*, the slope would increase somewhat more. Recall from your study of the part of this text entitled "Logarithms and the *D*–log *H* curve" (pp.) that the photographer moves the subject exposures laterally on the log *H* axis by changing the camera settings (aperture and shutter time). Increasing either the aperture size or the time of exposure moves the subject exposures laterally to the __________ .

 right

35. Conversely, closing down the aperture, or reducing the time of exposure, moves the subject exposures to the __________ .

 left

36. In the *toe* of the curve, because of the changing slope, a lateral movement of the subject exposures changes the contrast in the image. In the toe the slope increases to the right, so moving the subject exposures to the right __________ the image contrast.
 increases, decreases

 increases

37. For exposures in the toe, moving the subject exposures to the left __________ the image contrast.

 decreases

38. On the other hand, because the straight line is the region of *constant* slope, moving the subject exposures right or left *within* the straight-line region has ______
 no, a
 __________ effect on the image contrast.
 small, a great

 no

SELF-TEST ON EVALUATION OF THE NEGATIVE *D*–LOG *H* CURVE

Check your understanding of this chapter by answering the following questions. The correct answers follow the questions.

1. The steepness of the *D*–log *H* curve is numerically expressed by the __________ of the curve.
2. For a given subject, if the slope of the curve is great, the image contrast is __________ .
3. For a given subject, if the slope of the curve is small, the image contrast is __________ .

For the remaining questions, refer to the *D*–log *H* curve that follows.

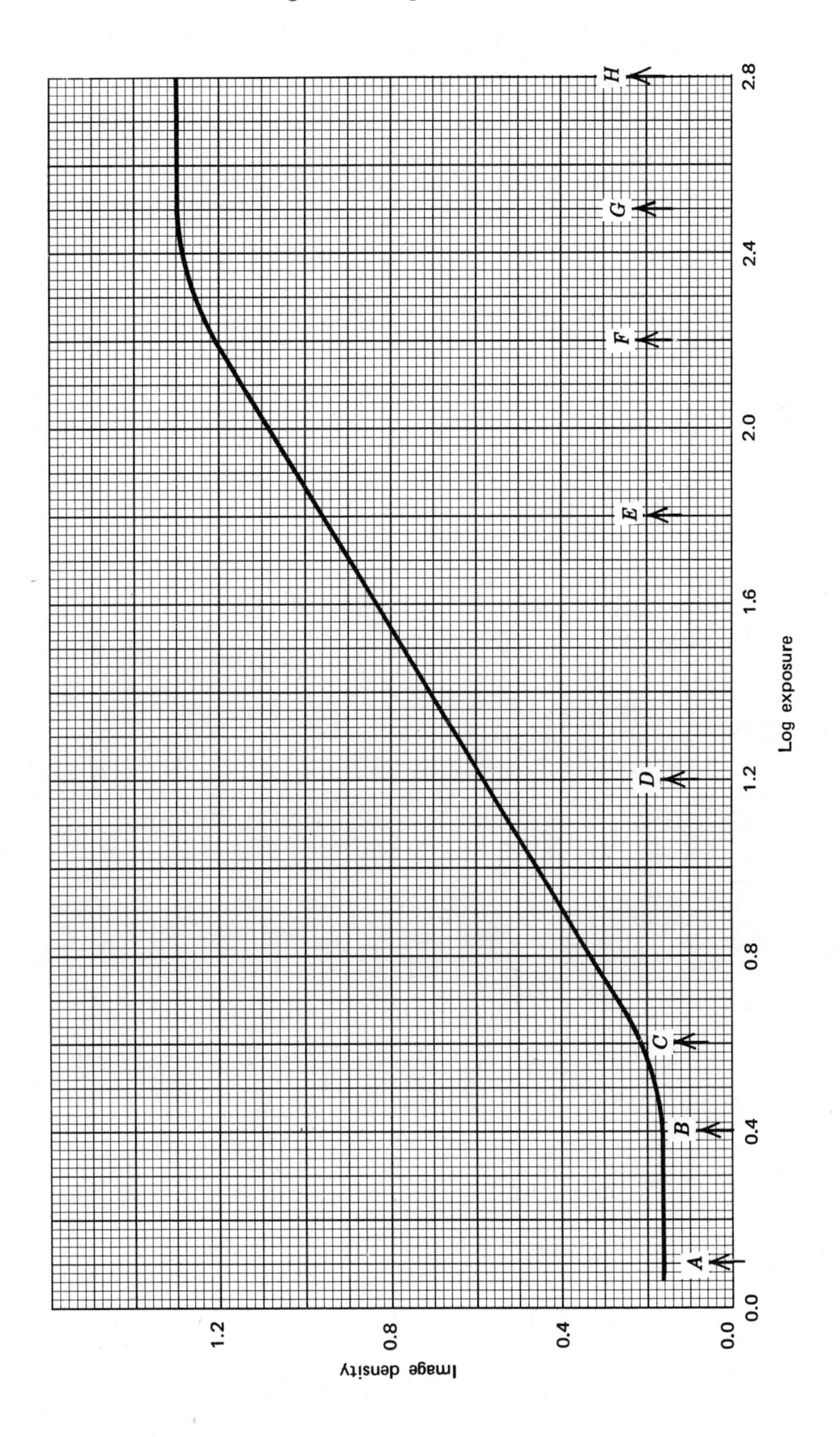

Image density
0.0
0.4
0.8
1.2
Log exposure
0.0
0.4
0.8
1.2
1.6
2.0
2.4
2.8
A
B
C
D
E
F
G
H

4. Between points *A* and *B* the slope is __________ .
5. For subject exposures lying between *A* and *B* the image contrast is __________ .
6. Between points *B* and *C* from left to right the slope __________ .
7. The numerical value of the slope between *C* and *D* is __________ .
8. As compared with the slope between *C* and *D*, the slope between points *D* and *E* and *E* and *F* is __________ .
9. As compared with the subject contrast (log *H* difference) the image contrast for the region between *C* and *F* would be ______________ .
   less, more, the same
10. Between points *G* and *H* the slope is __________ .
11. For this curve, the greatest image contrast for a given subject would be produced with exposures lying between points __________ and __________ .

ANSWERS TO SELF-TEST ON EVALUATION OF THE NEGATIVE *D*–LOG *H* CURVE

In the parentheses after each answer you find the numbers of the items in this chapter iv that relate to the question and its answer.

1. Slope (item 3).
2. Large, or great (item 1).
3. Small (item 2).
4. Zero (items 16 and 17).
5. Zero, or absent (item 18).
6. Increases (item 28).
7. 0.6 (items 4–10).
8. The same (item 13).
9. Less (item 2)
10. Zero (items 19–20).
11. *C* and *F* (item 21).

CHAPTER 5
Negative Contrast and Negative Exposure

In Chapter 4 you saw that: (a) the subject contrast is related to the differences in log exposure, (b) the image contrast is related to negative density differences, and (c) the relation between these two contrasts is given by the slope. This chapter involves the effects of the camera exposure level (*f*-number and time) and the subject itself on the negative contrast.

WHAT YOU WILL LEARN

If you work carefully with this chapter of programmed material, you will be able to:

1. Apply the concept of slopes to the reproduction of contrast (detail) to different tones of the subject;
2. Predict the change in image contrast as the exposure level in the camera is changed;
3. Identify underexposed and overexposed negatives by their relationship to the *D*–log *H* curve.

We now define a well-exposed negative. It is one in which there is sufficient contrast—detail—in every different area of the negative image, that is, for the shadows as well as the midtones and the highlights.

Consider first the situation sketched in the following graph. We show a specific *D*–log *H* curve, implying a specific photographic film material and a specific development process.

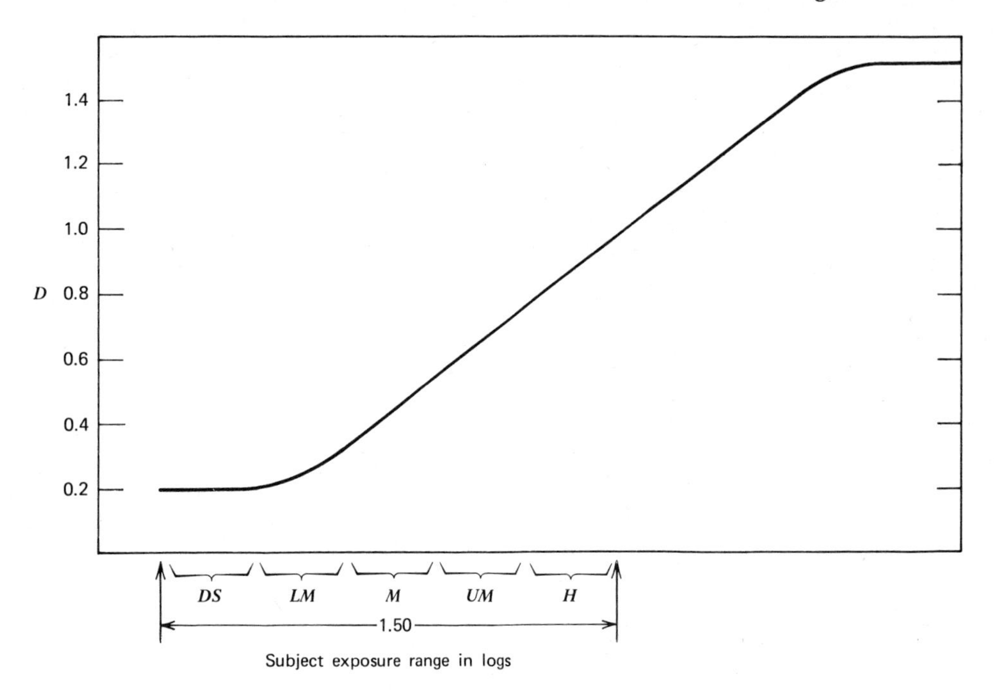

We assume in the sketch a subject that produces a total range of 1.50 in log exposure and that the camera settings (shutter and *f*-number) were such as to cause the exposures to lie on the log-exposure axis in the manner shown.

DIRECTIONS. Cover about 2 inches of the right-hand margin of each of the following pages in turn with an opaque sheet of paper. Read carefully the first numbered statement. *Write* the word or phrase that you believe correctly completes the statement. Move the cover sheet down to reveal the correct answer in the margin. Continue in this way until you have finished the section. Work at a rate that is comfortable for you, and no longer than you find pleasant. Begin when you wish.

1. You recall that one zone in a subject is equal to 0.30 in logs, and therefore this subject has a range of __________ zones.

   5

2. You recall further that the typical outdoor subject has a total range of about 7 zones, or __________ in logs.

   2.1

3. Thus we are assuming here that the contrast of the subject is somewhat __________ than normal.

   less

4. Different parts of the subject are identified with brackets and letters. The letters *DS* represent the "darkest shadow" tones of the subject. These subject tones fall on a part of the curve where the slope is practically __________ .

   0

5. Since the slope is practically zero for the part of the curve where the darkest shadows fall, the contrast in the negative for these subject tones would be __________ .

   0

6. If you were to look at this negative of this subject, you would see that the darkest shadows were blank and indistinguishable from each other. Furthermore, you would not be able to make a print in which the darkest shadows would have any detail whatever. On the other hand, at the right-hand end of the subject range, the bracket *H* represents the highlight areas; they fall on the straight line of the *D*–log

H curve. In the negative, the highlights would have maximum possible contrast, because the straight line is the part of the curve having the greatest ____________. — slope

7. To the left of the highlight area, the upper midtones *(UM)* and the midtones *(M)* are also exposed on the straight line; therefore the contrast of all of these tones would be ____________ that of the highlights. — equal to
   (more than, equal to, less than)

8. Because the midtones, the upper midtones, and the highlights all fall on the straight line of the *D*–log *H* curve, they would all be reproduced in the negative with maximum possible detail. The lower midtones *(LM)* fall on the *toe* of the curve. Since the slope in the toe is more than zero, but *less than* that of the straight line, the lower midtones would be reproduced with detail ____________ that of the midtones *M*. — less than
   (less than, equal to, more than)

9. The lower midtones would have at least *some* detail, although the detail for these tones would be weaker than that for the lighter midtones and the highlights. To summarize, this subject, so exposed, would have no detail for the darkest shadows, some detail for the lower midtones, and maximum detail for the midtones, upper midtones, and highlights. This negative was *underexposed;* its most obvious defect is the lack of detail for the ____________ ____________ of the subject. — darkest shadows
   (two words)

10. To obtain a better negative of this subject, it is necessary to increase the exposure level in the camera. We show in the following sketch the results of such an increase. We represent the same subject, the same photographic material, and the same processing, but exposed 1 stop more than in the first situation. The entire set of subject exposures is moved to the ____________, a distance of ____________ in logs. — right, 0.30

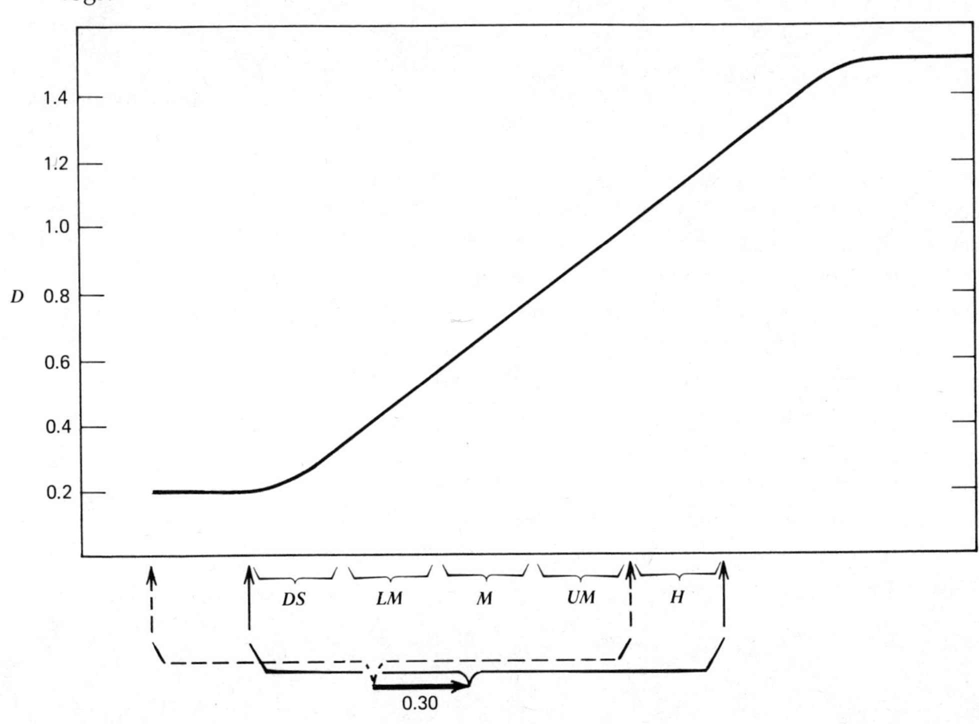

To show the change in exposure level (one stop), we show the *same D*–log *H* curve, in the *same* location on the log *H* axis. The curve does *not* change. What does change is the *location* of the subject exposures; they are all moved by 0.30 to the right, as indicated by the heavy arrow below the large brackets. The dotted bracket indicates the original location of the subject exposures; the solid bracket shows the new location of the set of subject exposures.

11. The deepest shadow areas in the first negative had no detail whatsoever. These tones *now* lie in the __________ of the *D*–log *H* curve.

toe

12. By the one-stop increase in camera-exposure level, the deepest shadows have been moved from the utterly flat part of the curve, where the slope is zero, into a region of the curve with at least some slope greater than zero. As a result, the change in camera settings has caused the contrast in the deepest shadows to __________ .

increase

13. At the other end of the subject exposure range, the highlight tones (which were on the straight line before) still fall on the straight line. The highlights will therefore have contrast __________ in the first negative.
less than, the same as, more than

the same as

14. The exposure change has had no effect on the highlight detail. Similarly, the upper midtones and the midtones have been moved to the right, but they still lie on the straight line, and their contrast is __________ as compared with the first case.

the same

15. Finally, the lower midtones have been moved to the right from the toe into the steeper straight line, and their contrast therefore is slightly __________ .

more

16. To summarize the effects of the one-stop exposure increase, no change has occurred in the contrast of the __________ , the __________ , and the __________ .

highlights
upper midtones
midtones (any order)

A slight increase in contrast has occurred for the __________ .

lower midtones

The really important change in contrast (from zero to at least *some* detail) has occurred for the __________ .

darkest shadows

You have seen that in an *underexposed* negative the problem is the lack of detail in the darkest shadows. You have seen that an increase in camera exposure moves the entire set of subject exposures to the right, thus placing the deepest shadows in a portion of the curve where some detail results. You have seen that for this somewhat flat subject (that produces a range of only 1.50 in log exposure) the one-stop increase in log exposure has some effect on the contrast of the lower midtones as they also are moved to the right. You have seen that the contrast of the rest of the subject tones is unaffected by the exposure increase, because those tones remain on the straight line.

17. We continue now to simulate graphically the production of yet a third negative of the same subject, with *2 stops more* camera exposure than in the second case. All the subject exposures will now be moved to the __________ a distance of __________ in logs, as in the following sketch.

right
0.60

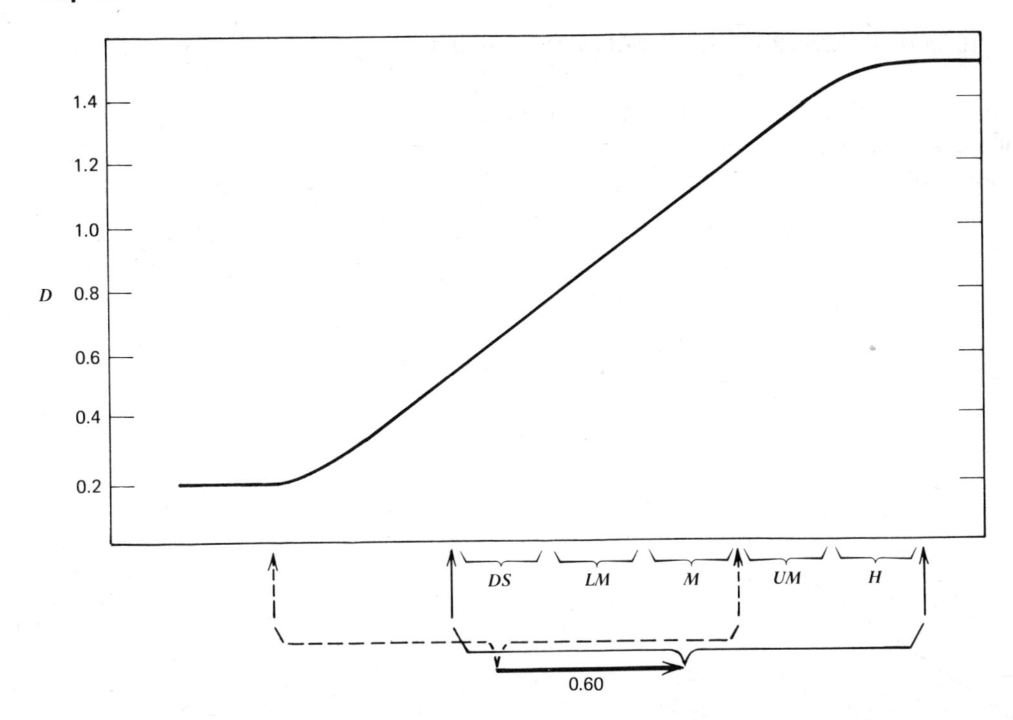

Note that the curve is the *same* as before; *the curve does not move.*

18. In the preceding sketch, the dotted bracket represents the exposure situation for the second negative; and the solid bracket, the third situation, made with 2 stops more camera exposure than the second. The darkest shadows have been moved again to the right, from the toe into the __________ __________. (two words)

    straight line

19. As compared with the second situation, the detail in the deepest shadows is now __________. less, the same, more

    more

20. The detail (contrast) in the deepest shadows is increased in the third negative because they have been moved into the straight line, the region of maximum slope. The midtones have also been moved to the right, but they still fall in the straight line (as they did in the second negative), so their contrast __________. increases, decreases, stays the same

    stays the same

21. On the other hand, the extreme highlights now fall far up on the shoulder of the curve, where the slope is practically __________.

    zero

22. Because the slope is practically zero on this part of the curve, the contrast of the highlights will be __________. great, moderate, almost zero

    almost zero

23. This third situation represents an *overexposed* negative. The negative has detail in all parts, except for the extreme __________.

    highlights

SELF-TEST CHAPTER V NEGATIVE CONTRAST AND NEGATIVE EXPOSURE

Check your understanding of this chapter by answering the following questions. The correct answers follow the questions.
Refer to the *D*–log *H* curve below.

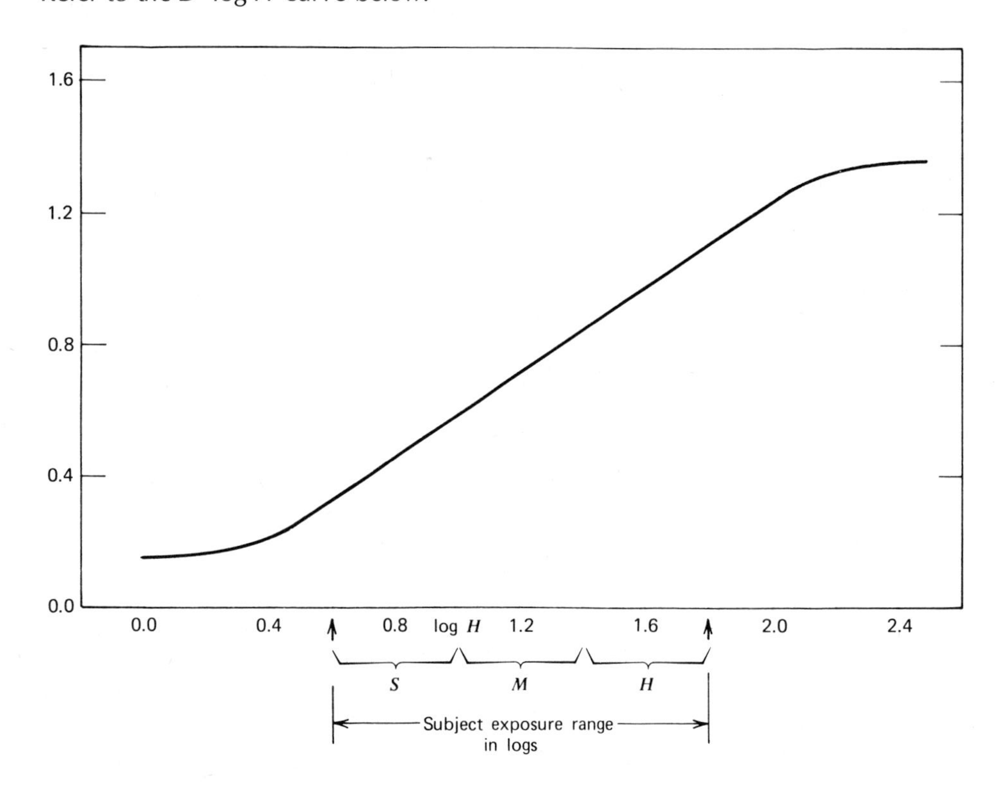

Note the range of the subject log exposures, and also the brackets indicating the Shadows, the Midtones and the Highlight subject exposures.

1. The subject range is ____________ zones.
2. Such a range represents a ____________________ subject.
   normal, flat, contrasty
3. The subject log exposures lie entirely in the __________ of the *D*–log *H* curve.
4. Because all subject exposures fall as shown, the contrasts of the Shadows, Midtones and Highlights are ____________ .
5. If a second negative were made of the same subject with *increased* exposure in the camera, ______________________ would move to the ____________ .
   the curve, the exposures
6. If the change in camera exposure level were two stops, the amount of shift would be ____________ in logs.
7. As a result of such a change only the contrast of the ____________ tones of the subject would change in the image.
8. The contrast of these tones would become ____________ .
9. Refer back to the original situation, but suppose that a third negative has been made with two stops *less* exposure in the camera. The only subject areas that would change in contrast from the original would be the ______ .

10. The effect on these tones would be __________ contrast than in the original negative.
11. In the third negative the contrasts of the Midtones and the Highlights would be __________ as compared with the original.

ANSWERS TO SELF-TEST ON NEGATIVE CONTRAST AND NEGATIVE EXPOSURE

In the parentheses after each answer you find the numbers of the items in this chapter that relate to each question and its answer.

1. Four (item 1).
2. Flat (items 2, 3).
3. Straight line (item 6).
4. Equal, or the same (item 7).
5. The exposures (item 10).
6. 0.60 (item 17).
7. Highlight (items 20–22).
8. Less, or smaller (items 21, 22).
9. Shadows (converse of items 11–14).
10. Less (converse of items 11–14).
11. The same, or unchanged (items 13, 14).

CHAPTER 6
Exposure Latitude

An important characteristic of a negative photographic material is the range of log exposures over which it produces a good image. This characteristic is associated with the range of subject tones that it can record and with the tolerance in camera settings that can be used.

If the usable range of log exposure values is small, the material can be used only with relatively flat subjects, and the camera settings must be nearly exactly correct. If the material has a wide exposure latitude, it can record subjects of varying tonal range and can allow unavoidable exposure errors.

Related to exposure latitude is the range of densities that the photographic material can produce. If this range is small, it limits the recording capacity of the material.

In the following we intend to define the usable limits of the *D*–log *H* curve as suggested by the sketch that follows. The problem is to specify the limiting points in the toe and the shoulder of the curve, that is, to identify the minimum and maximum *useful* log exposure values on the horizontal axis.

WHAT YOU WILL LEARN

If you work carefully with the following material, you will be able to:

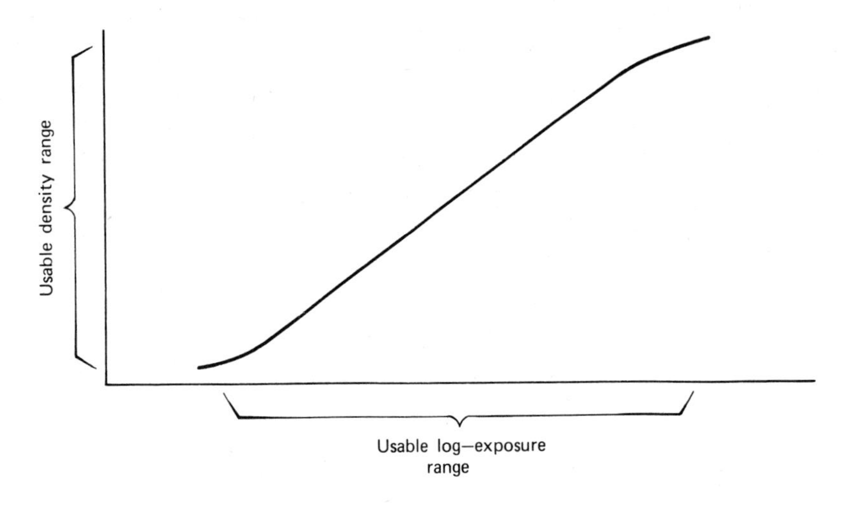

1. Locate the extreme useful points on both axes of the *D*–log *H* curve;
2. From these points, determine the range of log-exposure values that a film can record;
3. From the same points, determine the range of density values that a film can produce;
4. From the same points, and from knowledge of the range of subject exposures, find the tolerance in camera exposure that is available.

DIRECTIONS. Cover about 2 inches of the right-hand margin of each of the following pages in turn with an opaque sheet of paper. Read carefully the first numbered statement. *Write* the word or phrase that you believe correctly completes the statement. Move the cover sheet down to reveal the correct answer in the margin. Continue in this way until you have finished the section. Work at a rate that is comfortable for you, and no longer than you find pleasant. Begin when you wish.

1. From Chapter 5 you learned that detail in the negative is indicated by the ____________ of the *D*–log *H* curve at a given point. — slope
2. There you also learned that where the curve is horizontal, the slope is ____________ . — zero
3. Any part of the subject exposed on the horizontal part of the *D*–log *H* curve will have ____________ detail. — no
4. For the image to have sufficient detail, all the subject exposures must fall on a part of the curve having enough ____________ . — slope
5. Thus the problem of identifying the minimum *useful* log exposure value becomes the problem of knowing what is the minimum necessary slope. Extensive experiments have shown that for normal subjects, usual films, and normal development, the minimum slope is about *0.2* for recording shadow detail. The point having this slope is marked on the curve drawn in the following graph.

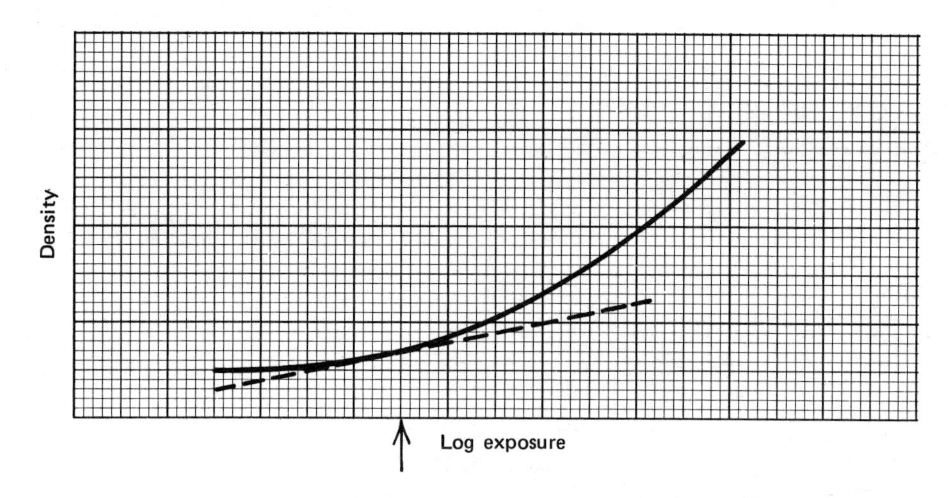

The dotted line is tangent to the curve at the point marked by the arrow. The slope of the tangent is 0.2, as is also the slope of the curve at that point. The arrow marks the minimum *useful* log exposure value. If any subject exposure lies to the left of this point, it will be reproduced with detail that is too ___________ .

weak, little, etc.

6. Subject exposures lying to the right of the minimum point (the arrow position in the graph above) will be reproduced with contrast that is ___________ than that at the arrow.

greater

7. Since the arrow marks the *minimum* useful slope, subject exposures to the right of the arrow will certainly have satisfactory contrast. At the other end of the *D*–log *H* curve, however, lies the extreme shoulder, also having a horizontal portion in the classical case, as shown in the part of a *D*–log *H* curve that follows.

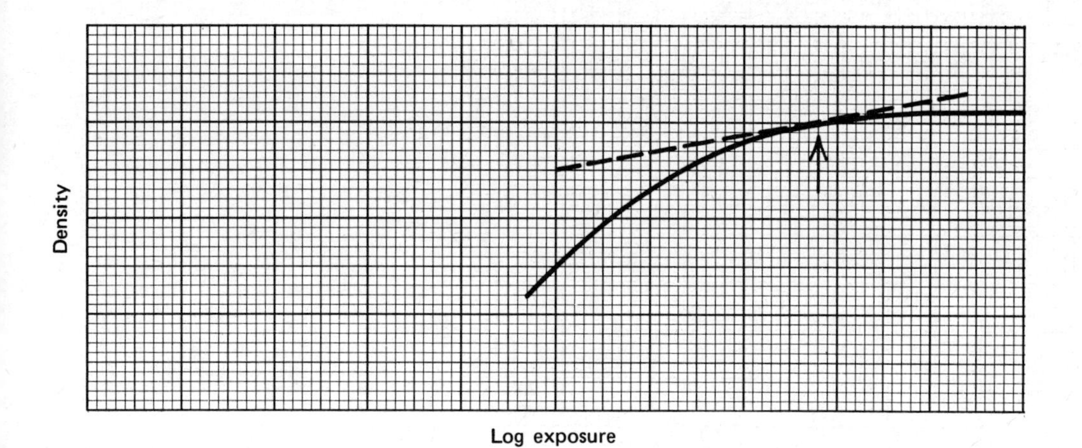

A dotted line with slope of 0.2 is drawn tangent to the curve; it marks the probable location of the *maximum* useful ___________ ___________.
(two words)

log exposure

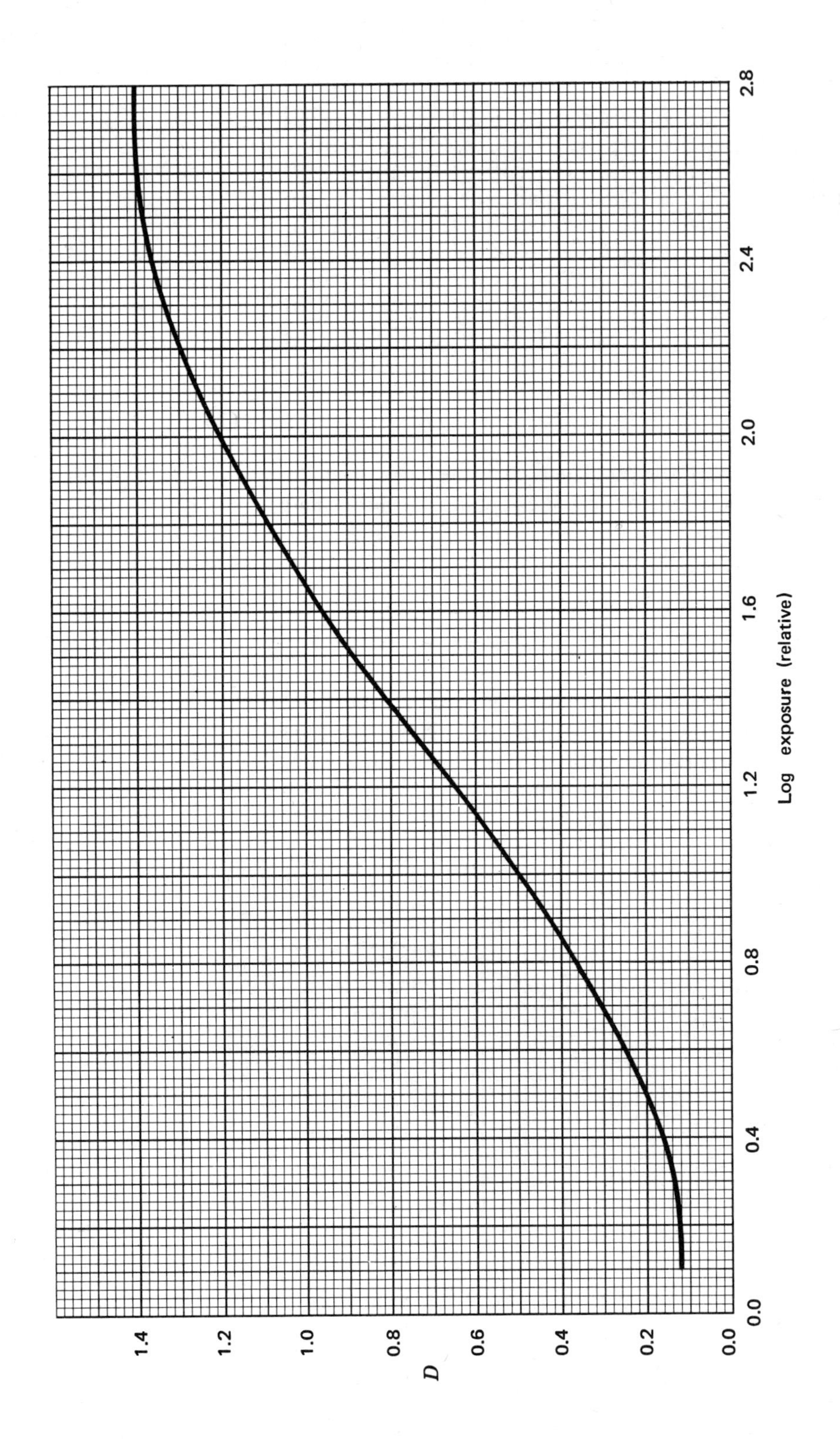
D
0.0
0.2
0.4
0.6
0.8
1.0
1.2
1.4
0.0
0.4
0.8
1.2
1.6
2.0
2.4
2.8
Log exposure (relative)

8. We have now identified the minimum and maximum useful points on the D–$\log H$ curve. In exposing a subject on the curve, no subject tone (the shadows on the left and the highlights on the right) should lie beyond these points. We define the total latitude of a negative material as the *horizontal distance* between the two useful points. It is a trial-and-error process to locate the useful points. An aid is a triangle made of thin card, with base 1.0 log units and height of 0.2 log units:

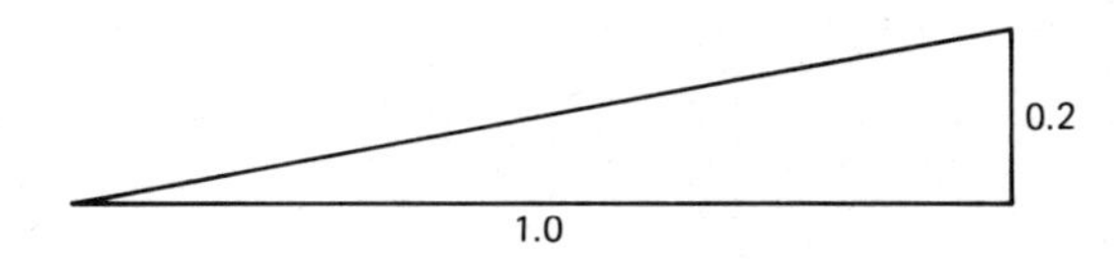

Make such a triangle and use it with the curve on page 74. Place it below the toe of the curve with the base aligned with the horizontal lines of the graph paper. Move it upward until it just touches the toe. The arrangement will resemble that shown below:

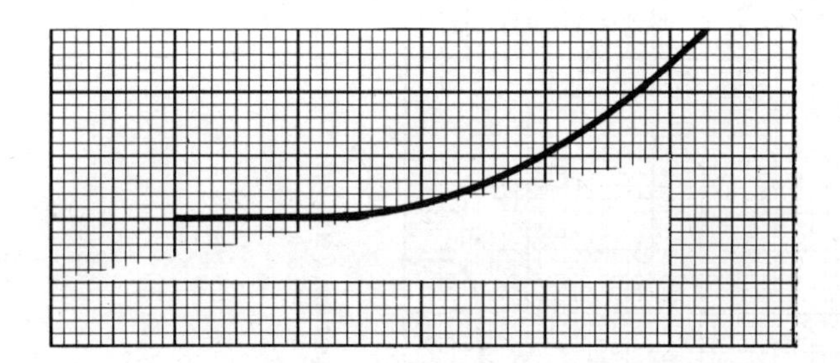

Estimate the log exposure value at the tangent point; the value is about ________ . (This estimate is hard; a difference of a few hundredths is of no practical importance.)

0.25

9. Turn the triangle around and use it similarly to find the tangent point in the shoulder:

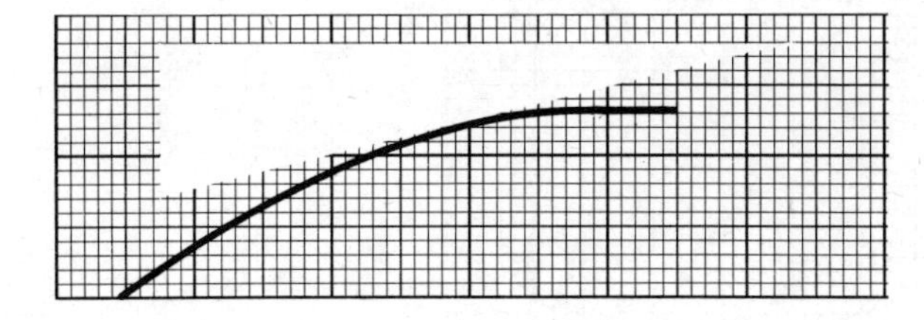

The log exposure value is ________ .

2.5

10. You have found in items 8 and 9 the log-exposure values for the minimum and maximum useful points. The difference in these values is the total latitude of the negative material. The difference is ________ .

2.25

11. Rounded off to the nearest tenth, the total latitude of this film is 2.2. The range of subject exposures in logs that could just be recorded by this film is also ________ .

2.2

12. The antilog of 2.2 is the maximum possible luminance range of the subject that this material could record. This range is ________ to 1.

160

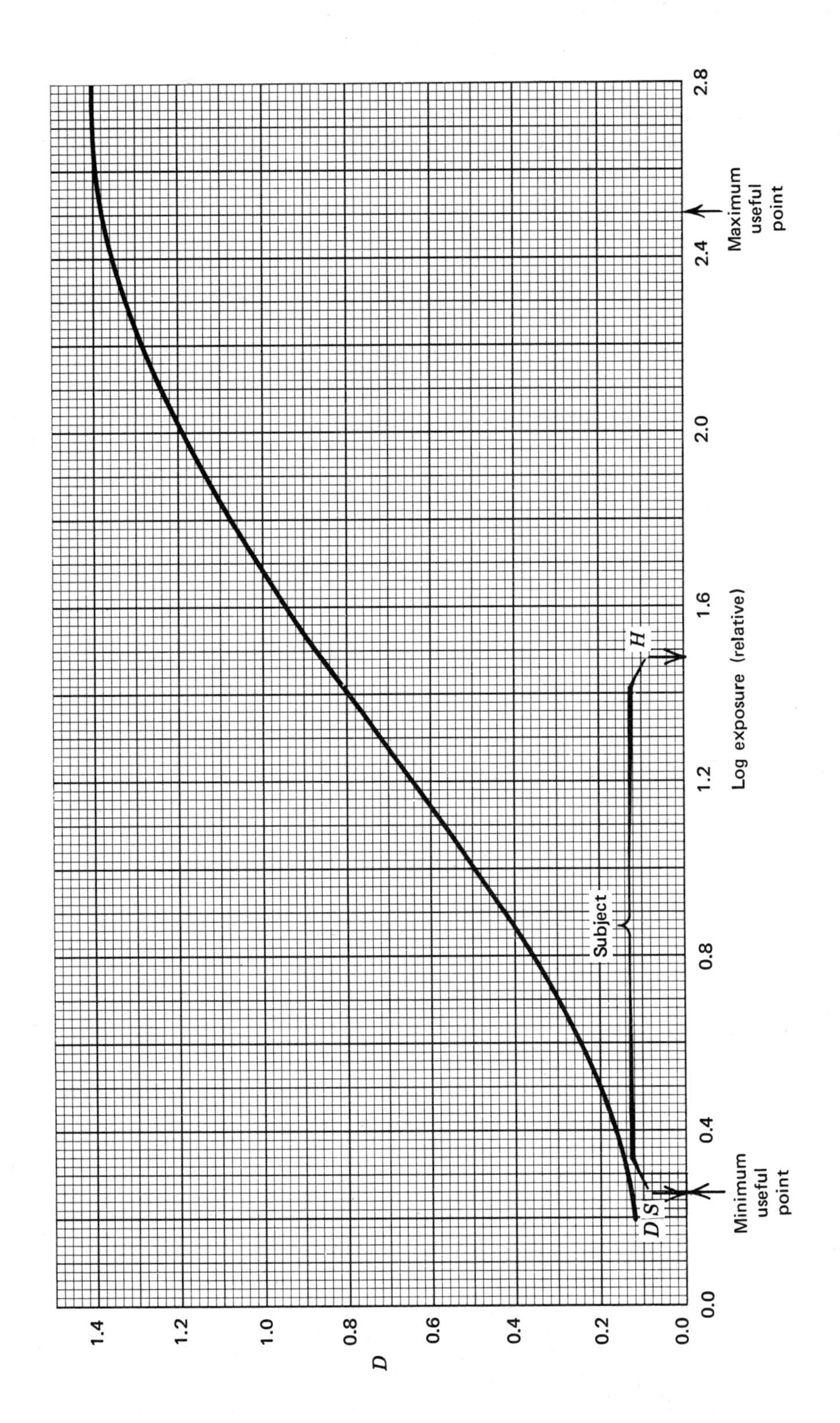
D
1.4
1.2
1.0
0.8
0.6
0.4
0.2
0.0
0.0
0.4
0.8
1.2
1.6
2.0
2.4
2.8
Log exposure (relative)
Minimum useful point
Maximum useful point
Subject
H
S
D

13. Because in the zone system one zone interval is in logs 0.3, this negative material could record a subject having a range of __________ zones. — 7+
14. If this material were to be used to photograph a subject of range greater than 2.2 in logs, it would be impossible to expose this subject so as to retain both shadow and highlight detail in the same negative. If the subject instead had a *smaller* range, some variation in exposure level would be possible. Assume a flat subject, with a range in zones of only 4, that is, a log-exposure range of __________ . — 1.2
15. The sketch on page 76 shows the extreme permissible leftward position of the subject exposures on the log H axis; the darkest shadow falls just at the minimum useful exposure point.

    There is room on the log-exposure axis to the right of the brightest highlight to the maximum useful exposure point. This distance is __________. — 1.0
16. This 1.0 interval is equivalent to __________ stops. — 3+
17. The meaning of this 3-stop interval is this: for this subject (having only a 4-zone range) and for this negative material developed in a specific way, there would be an allowable change in camera settings of 3 stops, from the least to the most, with a retention in the negative of both highlight and shadow detail. This tolerance in stops is often called the "exposure latitude" or the "latitude in exposure." The exposure latitude is determined by the total latitude of the film and *also* by the range of the tones in the __________ . — subject
18. If the total latitude of the film is 3.0 in logs, and if the subject has a range of 6 zones (1.8 in logs), the difference in these two values is __________ in logs. — 1.2
19. This difference of 1.2 is in logs the tolerance from least to greatest exposure level. It is equal to __________ stops. — 4
20. If the subject had a greater range (i.e., if it were more contrasty) such as 8 zones (2.4 in logs) the exposure latitude would be only __________ stops. — 2
21. For the same film, having a total latitude of 3.0 in logs, used to photograph a subject having a range of 10 zones, there would be __________ latitude in exposure. With *no* latitude in exposure, to reproduce all of the tones of the subject with sufficient detail, the camera settings would have to be *exactly* correct. — no
22. "Normal" camera exposure settings are assumed to place the subject exposures slightly to the right of the minimum, as in the following sketch:

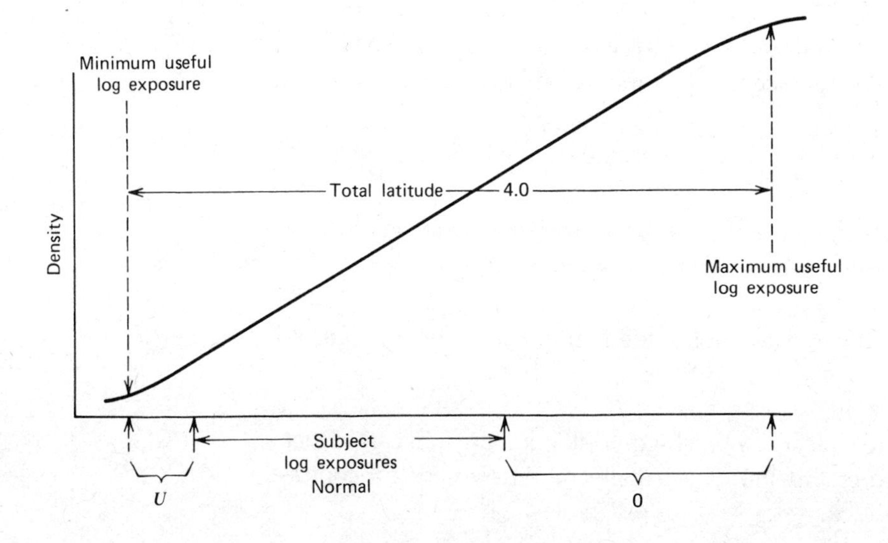

The bracket in the drawing marked *U* shows the latitude in exposure in the *under*exposure direction. If the bracket covers a distance of 0.3 in logs, the permissible latitude in this direction is __________ stop(s). 1

23. Similarly, the bracket marked *O* shows the latitude in exposure in the *over*exposure direction. Here the bracket would cover a distance of 1.8 (if the subject gave a log range of 1.9 in exposure) and the latitude in exposure in the overexposure direction would be __________ stop(s). 6
24. If the subject were "flatter" than the one in the example above (i.e., if it produced a smaller log-exposure range) the latitude in exposure would be __________ . (more, less) more
25. Conversely, if the subject were more contrasty (i.e., produced a greater range of log exposures), the latitude in exposure would become __________ . less
26. Having defined the minimum and maximum points on the *D*–log *H* curve, we can use these same points to estimate the range of density values that the photographic material can usefully produce. Please refer back to items 8–9, and read off the curve the density value for the minimum useful point. It is approximately __________ . 0.15
27. Similarly, read off the density value for the maximum useful point. It is about __________ . (Differences in these values of a few hundredths are trivial.) 1.37
28. The difference in the values found in items 26 and 27 is the range of useful densities of this material. The difference is __________ . 1.22
29. The meaning of the density range is in part that it measures the range of tones that the material can produce. The antilog of 1.22 is about __________ . 17

The number just found (17) means that this material could produce at most a tonal ratio of 17 to 1. This range of densities is very small, not typical of silver-halide films. Consider, in summary, the special-purpose "lith" film having the *D*–log *H* curve shown on page 79.
Films used for lithographic purposes are not ordinarily used pictorially, and therefore the 0.2 slope does not necessarily locate the minimum and maximum points. We have here arbitrarily used a density respectively 0.1 above and below the minimum and maximum densities to identify the useful points on the curve; they are labeled with arrows *A* and *B*.

30. The interval between the arrows on the *horizontal* axis shows the range of log exposures that the material can *accept*. This interval is about __________ in logs. 1.0
31. The antilog of 1.0 is the extreme tonal range that the material can *respond to*. It is about __________ to 1. 10
32. On the other hand, the interval between the arrow positions on the *vertical* axis shows the range of densities that the film can *produce*. This interval is about __________ . 2.9
33. The antilog of 2.9 shows the range of tones the film can usefully *produce*; it is __________ to 1. 800

The data in items 31–34 are typical of a "high-contrast" film; the film can *accept* a very small range of log exposures, thus the subject matter must have very low contrast,

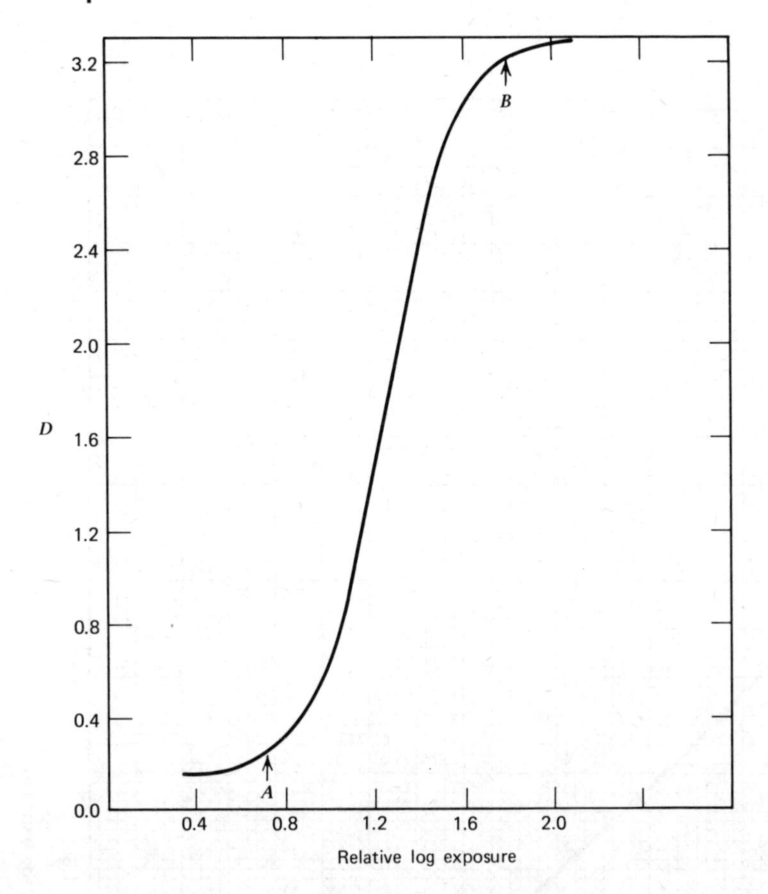

that is, be very "flat." Such film is typically used with uniformly lighted drawings, as of line copy, where the contrast of the subject is hardly more than 1.5 in logs. ("Black" ink on "white" paper really involves only two tones—one dark gray and one light gray.) Furthermore, because the useful range of log exposures is so small, the camera exposure level must be accurately controlled—the tolerance in exposure settings is almost nonexistent.

This film, however, *produces* a very great range of densities, and thus from a low-contrast original produces a high-contrast image.

SELF-TEST ON EXPOSURE LATITUDE

Check your understanding of this chapter by answering the following questions. The correct answers follow the questions. Please refer to the graph on the following page.

1. Locate the minimum useful point where the slope is 0.2.
   At this point the relative log exposure is ____________ .
   At the same point the density is ____________ .
2. Locate the maximum useful point. At this point, the relative log exposure is ____________ and the density is ____________ .
3. The total log exposure range between the two points is ____________ .
4. The total density range between the two points is ____________ .

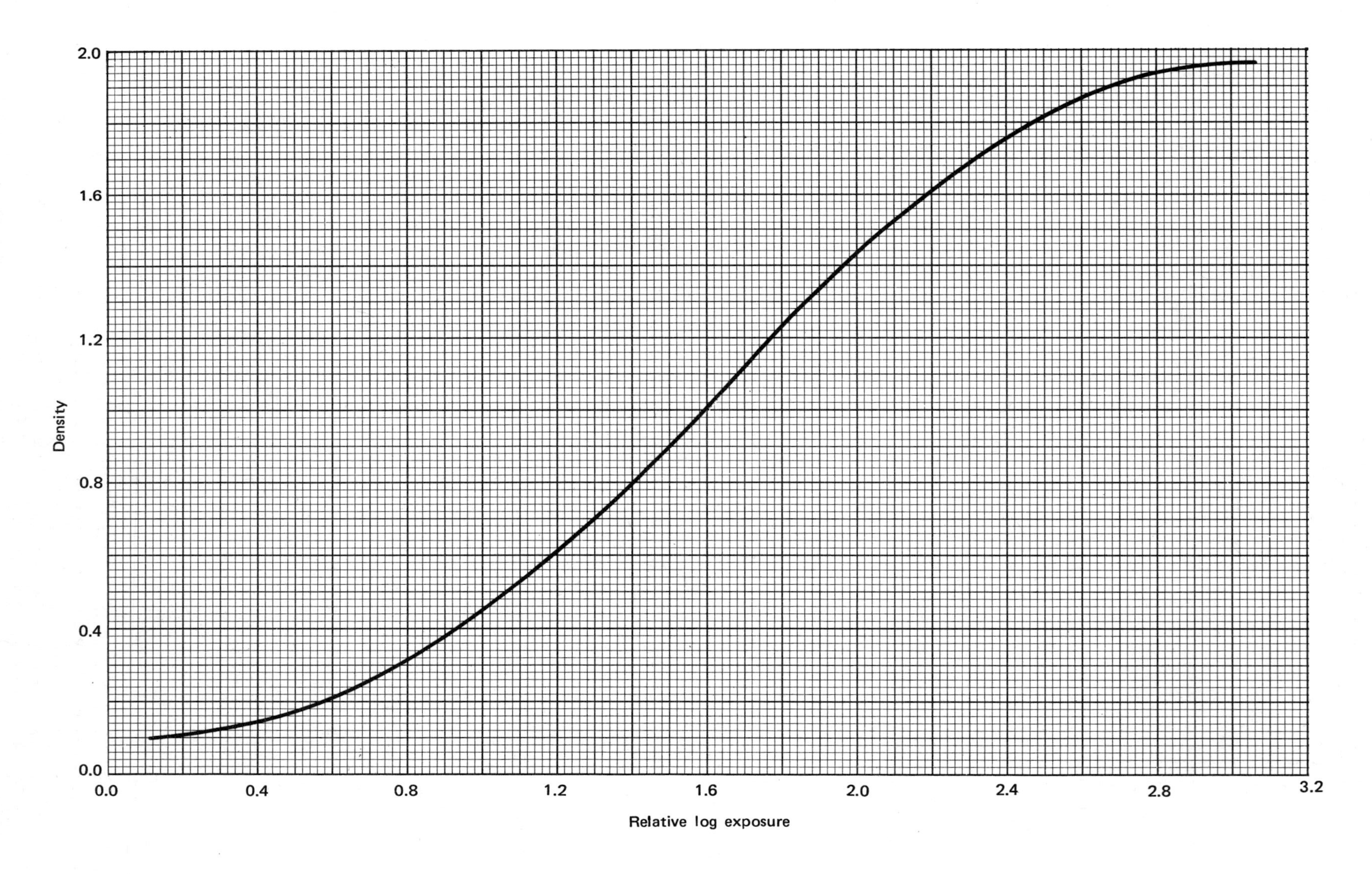
Density
2.0
1.6
1.2
0.8
0.4
0.0
0.0
0.4
0.8
1.2
1.6
2.0
2.4
2.8
3.2
Relative log exposure

5. The greatest possible range of subject exposures that could be recorded by this film-development process is (in logs) ___________ .
6. If the subject range of log exposures were only 1.9, there would be a total tolerance in log exposure (from least to most) of ___________ , or ___________ stops.
7. In general, as the subject log range (subject contrast) is reduced, the latitude is ___________ .
8. The range of tones this film-development process could produce is at most ___________ .

ANSWERS TO SELF-TEST ON EXPOSURE LATITUDE

In the parentheses following each answer you find the numbers of the items in this chapter that relate to the question and its answer.

(For the numerical answers, you are correct if you agree with those given below within a few hundredths in logs.)

1. 0.35 (items 5, 8); 0.13 (item 26).
2. 2.85 (items 7, 9); 1.93 (item 27).
3. 2.50 (item 10).
4. 1.80 (item 28).
5. 2.50 (item 11).
6. 0.60, 2 (items 15–19).
7. Increased, or greater (items 18–20).
8. 70 to 1 (item 29).

CHAPTER 7

Negative Development and Negative Contrast

In earlier sections you saw that:

1. Slope is a useful measure of the relationship between image contrast and the contrast of the subject.
2. Slope is related to such practical concepts as image detail and the rendition of tones.
3. When for a given subject the camera exposure settings (shutter time or *f*-number) are changed, the subject exposures are shifted laterally on the log-exposure axis of the *D*–log *H* curve.
4. Such a lateral shift usually moves the subject exposures to different positions on the *D*–log *H* curve. The result is to place the subject exposures in regions of different slope, and thus of different contrast.

We turn now to the effects on the image of a change in *development time* of the negative. These effects are shown by a set of *D*–log *H* curves produced with varying development time but with fixed conditions of exposure and of other factors in development (chemistry, agitation, and temperature). In the diagram on page 84 is a set of such curves for a generalized negative material. In the following, please refer to these curves.

Each of the *D*–log *H* curves has the classical shape: a horizontal portion at the extreme left for very small exposures; next a toe; next a straight line; finally a shoulder that is nearly flat at the extreme right for very great exposures.

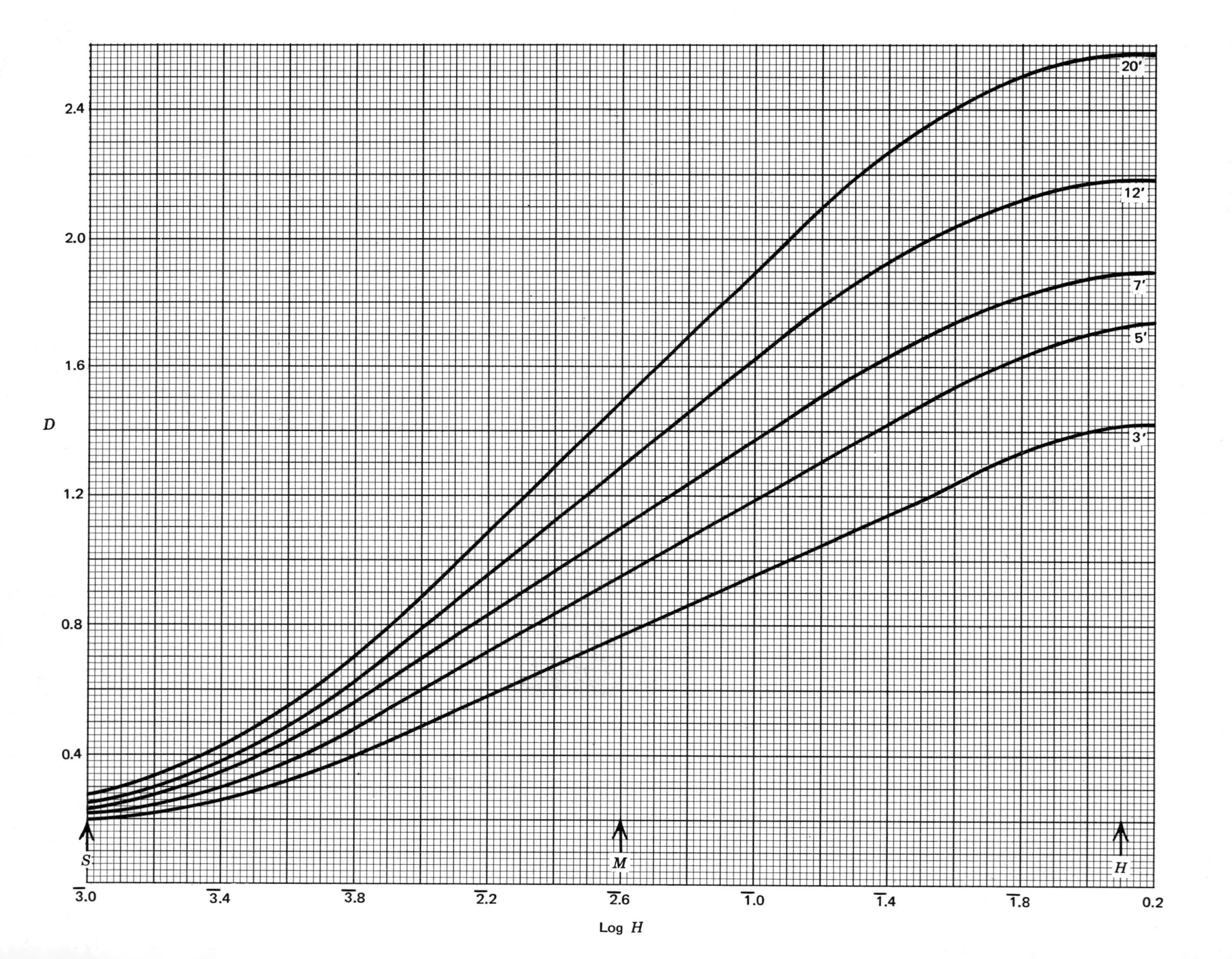
2.4
2.0
1.6
D
1.2
0.8
0.4
20′
12′
7′
5′
3′
S
M
H
$\bar{3}.0$
$\bar{3}.4$
$\bar{3}.8$
$\bar{2}.2$
$\bar{2}.6$
$\bar{1}.0$
$\bar{1}.4$
$\bar{1}.8$
0.2
Log H

WHAT YOU WILL LEARN

If you work carefully with the following material, you will:

1. See how differently-exposed areas of the negative change in density with time of development;
2. Be able to measure the extent of development of a negative;
3. Know how the contrast of different areas of the negative changes with development time.

DIRECTIONS. Cover about 2 inches of the right-hand margin of each of the following pages in turn with an opaque sheet of paper. Read carefully the first numbered statement. *Write* the word or phrase that you believe correctly completes the statement. Move the cover sheet down to reveal the correct answer in the margin. Continue in this way until you have finished the section. Work at a rate that is comfortable for you, and no longer than you find pleasant. Begin when you wish.

1. Inspection of the set of curves shows that as development time increases, each successive curve lies completely above the preceding one. Thus for every exposure the density becomes __________ as development proceeds. — greater, larger (etc.)
2. Although for every exposure the density increases with increasing development time, the densities for different exposures by no means increase by the same amount. At the extreme left of the curves, for example, the density produced after 3 minutes of development is very small—only __________ . — 0.20
3. After 20 minutes of development, as shown on the uppermost curve for the same log-exposure value, the density is still very small—only __________ . — 0.28
4. Thus although the development time has increased by over 8 times, for this exposure the density has increased from 0.20 to only 0.28. On the other hand, at the extreme right of the curves, the density is, after only 3 minutes of development, the much larger value of __________ . — 1.42
5. At the same log exposure value as in item 4, after 20 minutes of development time, the density has increased very greatly, to a value of __________ . — 2.57
6. Just these two examples show that the effect of increasing development time depends greatly on the exposure that the negative material has received. The greatest change in density with development is for the __________ exposure. — greatest
7. You can see the relationship between exposure level and the effect on density with development more clearly if you read off points from the set of curves at fixed exposure levels and then make new curves showing how the density changes with development time.

   Please do so first for the log-exposure value at the arrow marked *S*. The density values in order, curve by curve from the bottom, are: __________ , __________ , __________ , __________ , __________ . — 0.20, 0.22, 0.24, 0.26, 0.28
8. Now plot these densities against development time on the following sheet of graph paper. Draw in the line showing the growth of density for this exposure as development time increases.

   The line you have drawn is nearly straight and nearly horizontal, indicating that for this exposure value the density changes with development time only very __________ . — slowly, slightly (etc.)

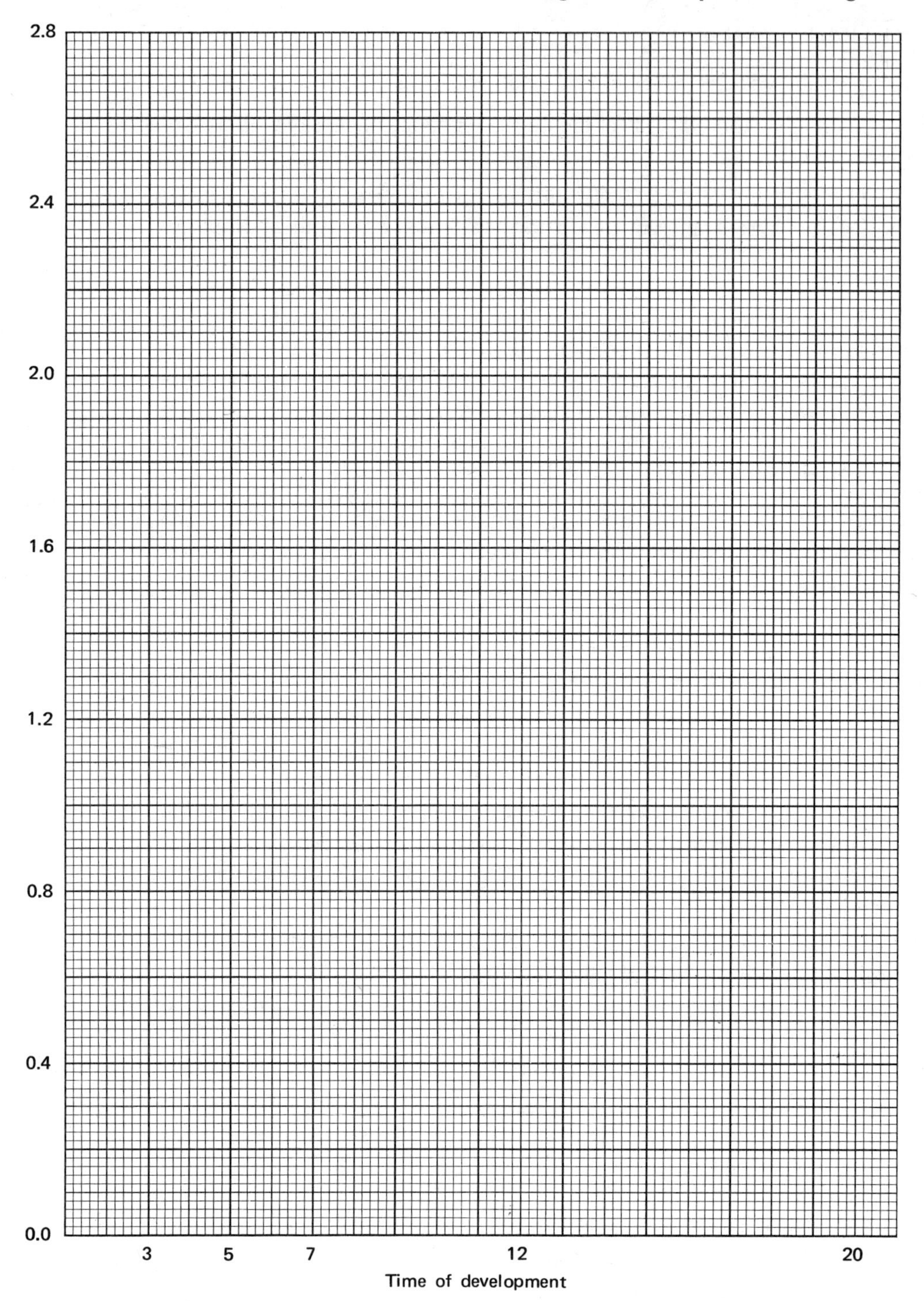

9. Now repeat the same process for the log-exposure value at the arrow labeled *H*. In order from the bottom curve upward, the density values are: ____________, ____________, ____________, ____________, ____________.

1.42, 1.72, 1.90, 2.18, 2.57

10. Please plot these values against development time on the same graph paper you used for item 8.

The line you have drawn is strongly curved. Also, it rises very rapidly, unlike the one you drew before. The rapid rise means that for this log-exposure value the density increases very __________ with increased development time.

greatly, or rapidly (etc.)

11. To complete the analysis in part, repeat the same process with an intermediate exposure level, at the arrow marked *M*. The density values in order are: __________, __________, __________, __________, __________.

0.76, 0.95, 1.10, 1.29, 1.49

12. The curve you have plotted shows a moderate rise in density with increased development time. You find the three curves correctly plotted on the graph that follows. In summary, if *S* represents a shadow tone in the subject, *M* a midtone,

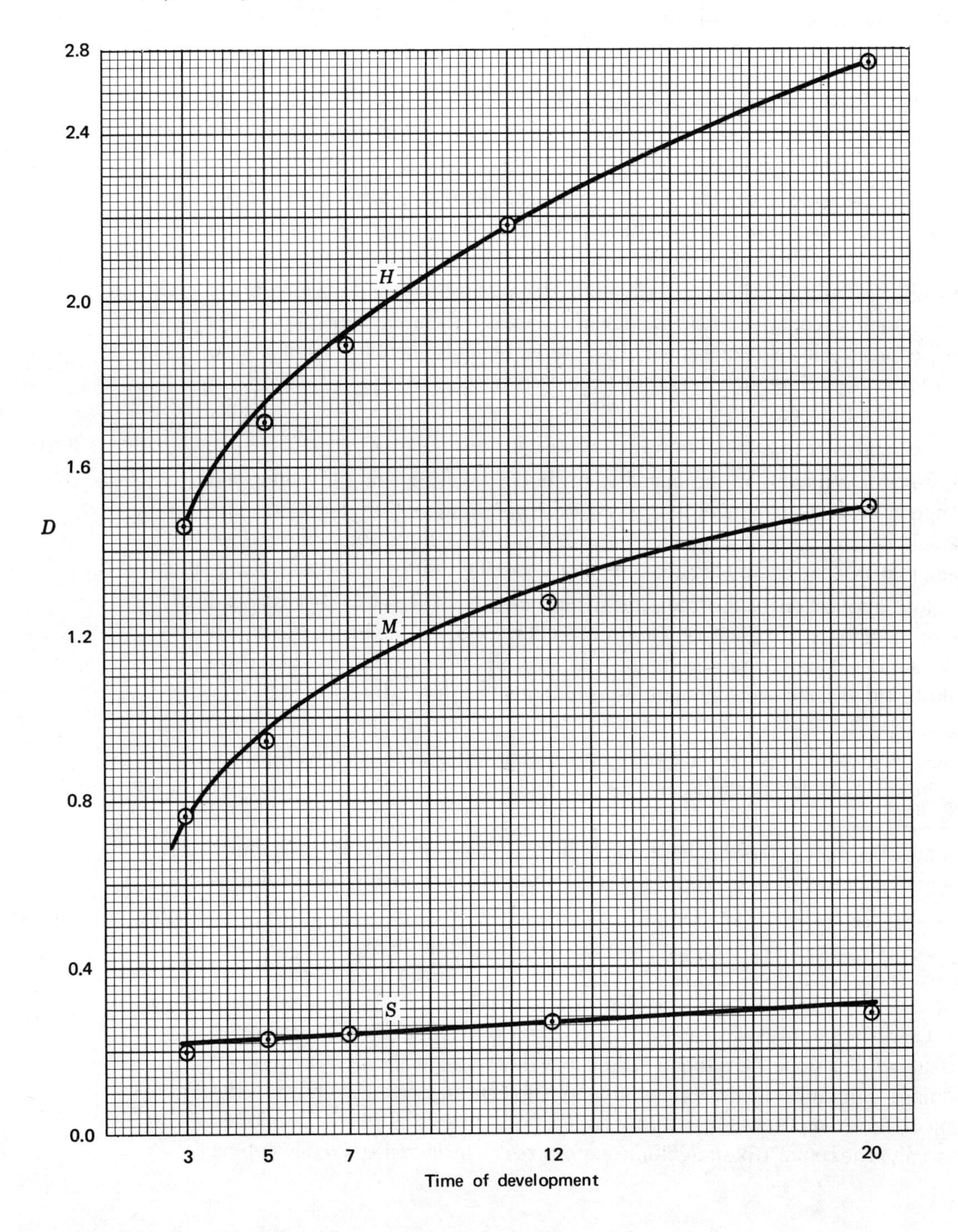

and H a highlight, what you have plotted shows that for exposures like these three falling on the D–log H curve, increased development time causes the greatest increase in density for the __________ . highlight

13. On the other hand, increased development time causes only a very small increase in density for the __________ exposure. shadow

14. The midtone density increases moderately with development. Because development affects each of these three tones differently, their relationship also changes as development goes on.

   Such a relationship is best shown by the change in *slope* of the D–log H curves. The most obvious change is in the slope of the __________ __________ of the curves. (two words) straight lines

15. The slope of the *straight-line* portions of the D–log H curves continuously increases with development time. Recall that slope is a measure of the __________ of the image. contrast

16. Since the slope of the straight line increases with increasing development time, image areas exposed on the straight line of the D–log H curve will increase in __________ as development goes on. contrast

17. The slope of the straight line is called *gamma*. It has long been used as a measure of "development contrast." Consistency in processing photographic images is shown by a stable value of gamma. If, however, replenishment is inadequate (so that for a given time the amount of development lessens), this is shown by a decrease in the measured value of __________ . gamma

18. Conversely, a rise in gamma shows an increase in the amount of __________ . development

19. Because of variations in the agitation method used, specifying the time and temperature of development in a given developer may not produce the same amount of development in a given film. For this reason, the amount of development may be specified by the attainment of a specific value of __________ . gamma

20. Another application of this measure of the amount of development is to an experiment to test, for example, the effect on film speed caused by the use of two different developers. The development time in the two developers should be adjusted to give the same amount of development as shown by equal values of __________ . gamma

21. Since the greatest possible value of gamma is mainly a property of the film itself, we identify a "high-contrast" film by its ability to produce a large value of __________ . gamma

22. On the other hand, "low-contrast" films, such as those intended for portraiture, are manufactured so as to produce at maximum only a relatively small value of __________ . gamma

23. Lithographic films can produce very large values of gamma, as much as 5 or even more. These are called __________ __________ films. (two words) high contrast

24. Most general-purpose films can be developed so as to produce various values of the straight-line slope. Development time may be adjusted (as in the zone system) to improve negative quality (printability) for subjects of different contrast. A flat subject, for example, will give a low-contrast negative if the film is developed normally. The contrast *for straight-line* exposures can be increased by developing

the film to increase the straight-line slope, that is, to a greater-than-normal value of ________ . — gamma

25. Conversely, for a contrasty subject, one may reduce development time below normal to obtain a lesser value of gamma and thus to reduce the contrast of the image exposed on the ________ ________ of the D–log H curve. (two words) — straight line

26. Good negatives of normal subjects are usually obtained by processing the negatives so as to obtain a gamma value of 0.6–0.7. Recall that a normal outdoor subject has a log range of luminances of about 2.1, about 7 zones. If a subject had a range of only 5 zones, that is, a log luminance of 1.5, the negative could probably be improved by processing to a ________ value of gamma. — high

---

27. You have seen that the slope of the *straight line* increases markedly with time of development. In other areas of the curve, however (the toe and the shoulder), the effect of increasing development time is different. As an extreme example, observe the toe areas of the set of curves at the beginning of this section near the arrow labeled *S*. For the bottom curve, representing 3 minutes of development, the slope is very small, almost zero. Even after 20 minutes of development (shown by the uppermost curve) the slope is still very ________ . — small

28. The preceding observation means that for subject exposures falling far to the left in the extreme toe of the curve, even prolonged development hardly changes the slope, and thus the image ________ . — contrast

29. The part of the curve near *S* is the region of ________ exposure. — under

30. In an underexposed negative, the extreme shadow areas have very little contrast in the negative, even if development time is very ________ . — long, great (etc.)

31. Meanwhile, as development proceeds and the shadow areas change hardly at all in contrast, the exposed areas near arrow *M* (which are on the straight line) are increasing in ________ . — contrast

32. To say that "development determines negative contrast" is certainly valid for those exposures that fall in the ________ ________ . (two words) — straight line

33. The same saying is not really valid for those exposures that fall far to the left in the ________ of the curves. — toe

34. Thus for a negative that has been grossly underexposed, even prolonged development will hardly improve the contrast of the extreme ________ areas of the image. — shadow

35. Now consider the shoulder regions of the curves, near the arrow marked *H*. After 3 minutes of development, the slope near *H* is very ________ . — small

36. A very small slope is an indication of almost no image ________ . — contrast, or detail

37. As shown by the uppermost curve, even after 20 minutes of development the slope in this same shoulder region is still very small, almost zero. Thus development even for this long time has had little effect on the image ________ in this region. — contrast

38. If any subject exposures fall far to the right, on the nearly flat part of the shoulder, a change in development time hardly affects the contrast (detail) for those image areas.

In summary, there is only *one* region of the curve where development strongly affects image contrast; this region is the ________ ________ . (two words) — straight line

39. There are two regions of the curve where a change in development time has little effect. These regions are the extreme portions of the ________ and the ________ . — toe, shoulder (either order)

40. The use of gamma as a measure of the amount of development of a negative material is gradually being abandoned in favor of a better measure. One reason for the change is that many modern negative materials have no straight line. If there is no straight line, it is impossible to find a value for ________ . — gamma

41. For example, of the following sketched *D*–log *H* curves, gamma could be found only for the curve labeled ________ . — *B*

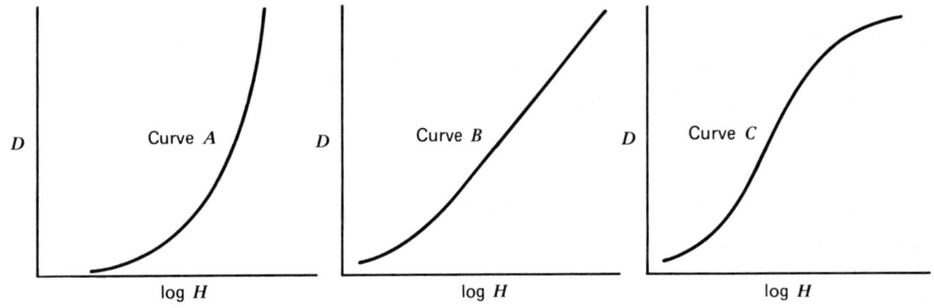

42. Other negative materials may have a straight line, but of insignificant length; still others have two straight lines of different slope. Gamma can certainly be found for these materials, but the value is not useful. In addition, there is good evidence that excellent negatives do not require exposures only on the straight line. In fact, best negative quality usually results when some of the shadow exposures lie to the left of the straight line, that is, in the ________ of the curve. — toe

43. For these reasons, among others, contrast index (CI) is replacing gamma. Contrast index is an *average slope* for that part of the curve that represents *excellent* negative exposures. The part of the curve that should be used was found through an extensive study of a large number of excellent negatives made of a variety of subjects with different films and development. To obtain a sufficiently good approximation to the contrast index, follow these steps, using the 3-minute curve on the set with which you have been working. First, note the base & fog level for the 3-minute curve. It is the density value for the flat part of the toe of the curve; it is ________ . — 0.20

44. Increase this value by 0.10, to get ________ . — 0.30

45. Look to the right on the 3-minute curve and place a tick mark on the curve at a density of 0.30. Now place a ruler so that the 0 mark is on the tick you made, and so that the 5-inch scale division of the ruler intersects the curve. The arrangement will look like this:

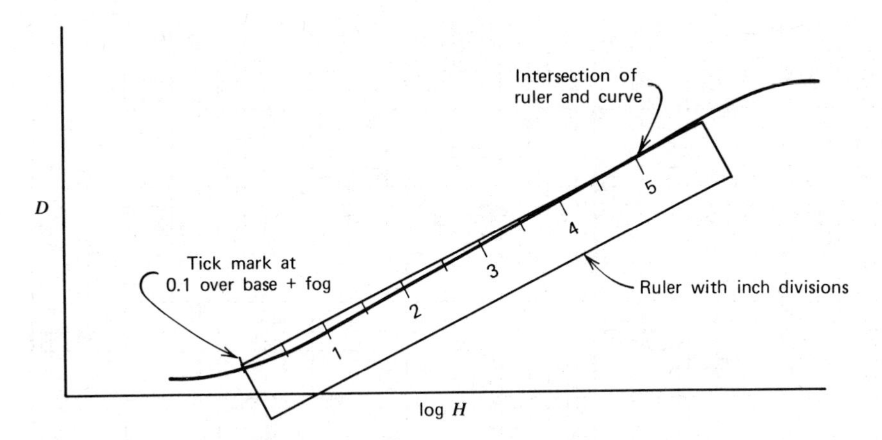

The contrast index is the slope of the straight line made by the edge of the ruler. Following the usual method, the slope is about ________ .

0.4; a variation of a few hundredths is trivial

46. The approximate method you have used is useful only for *D*–log *H* curves that have been developed to "normal" CI values, that is, over the range of curves in the set you have been using. For practice, use the same method to find the CI value for the 20-minute curve. The contrast index for that curve is about ________ .

0.90

47. When you find the CI value, you find the average slope over the most-used portion of the *D*–log *H* curve. Since CI *is* a slope, it is a measure of the contrast of the negative image as compared with the contrast of the ________ exposures.

subject

48. Also like gamma, CI changes with development conditions (like development time); it is a measure of the amount of ________ .

development

49. Therefore CI, like gamma, is used to measure development stability and to adjust development conditions to obtain a desired "development contrast."

A more precise way of finding *CI* is by the use of the transparent overlay in an envelope at the back of the book. The overlay has: (a) a large arc at the right, with numbered divisions, (b) a small arc at the left, with similar numbers, and (c) a horizontal straight line labeled "density of base plus fog density."

Place the overlay on the 3-minute curve at the beginning of this section as follows; Align the horizontal line on the flat part of the toe of the curve, making it parallel with the horizontal lines of the graph paper. Move the overlay *horizontally* until the *D*–log *H* curve intersects *both* arcs at the *same* scale division value. Now record that value, which is the CI: ________ .

0.42

50. Please use the CI meter on each of the curves of the set. In order from the bottom, the values are approximately: 3 minutes—0.42; 5 minutes— ________ ; 7 minutes— ________ ; 12 minutes— ________ ; 20 minutes— ________ . Again, variations of a few hundredths from these values are insignificant.

0.55; 0.64; 0.77; 0.91

51. In manufacturer's literature you will often see a graph of *CI* versus time of development. Please make such a graph using the following graph paper and the preceding data:

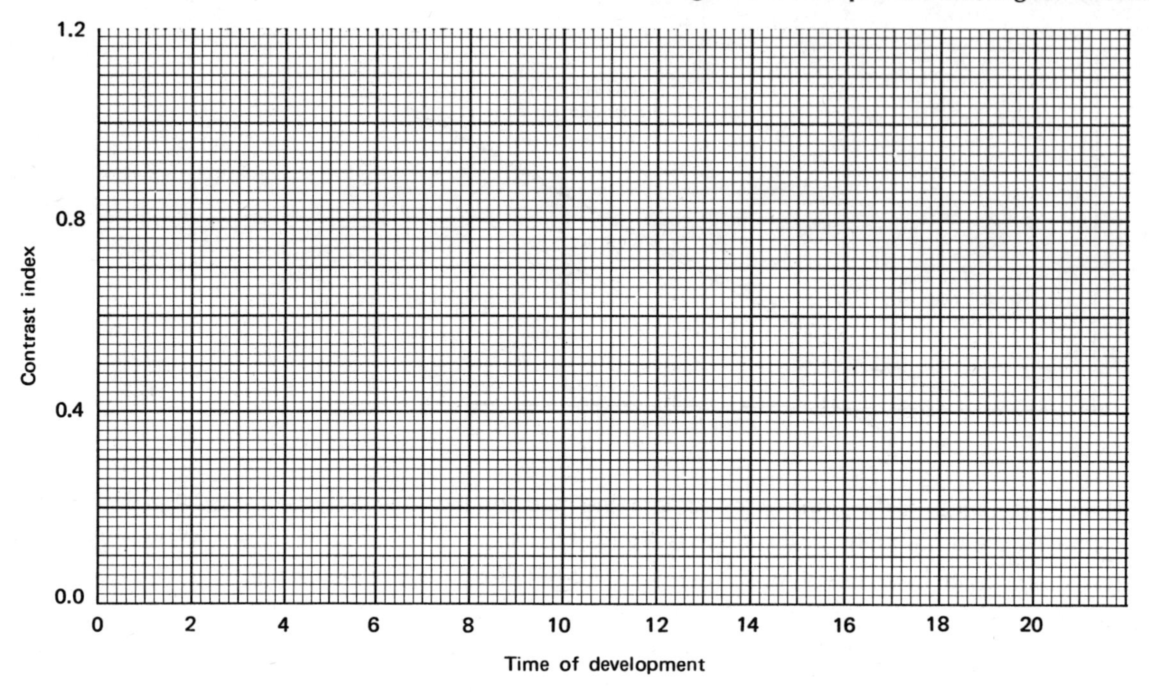

(A correct graph is shown below.) It rises rapidly at the left, showing that CI

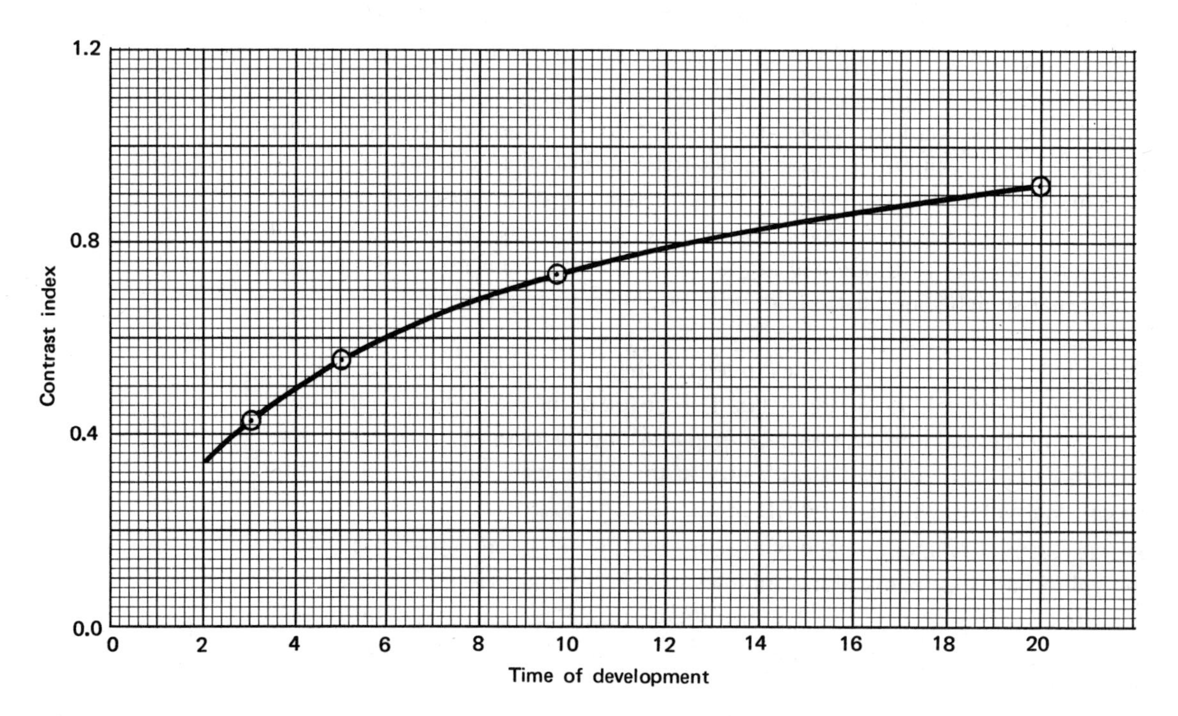

changes quickly in the early stage of development. The curve flattens out at the right, showing that CI changes only slowly with prolonged processing times. The graph suggests that there is an upper limit to the value of CI; this value is determined mainly by the type of film.

52. From a graph like the one above you may find the time of development needed to produce any desired CI. If, for example, a CI of 0.60 is needed, from the plot the necessary time of development is __________ minutes. 6

The graph is valid *only* for a specific set of development conditions, including chemistry, temperature, and the agitation method, as well as time of development. Published curves can be reliably useful only if your conditions agree in all essentials with those under which the data were obtained.

SELF-TEST ON NEGATIVE DEVELOPMENT AND NEGATIVE CONTRAST

Check your understanding of this chapter by answering the following questions. The correct answers are found after the questions.

1. As development proceeds, how do *densities* change for exposures lying in the:
   (a) toe of the *D*–log *H* curve

   ____ ;
   slight increase, moderate increase, great increase

   (b) straight line of the *D*–log *H* curve

   ____ ;
   slight increase, moderate increase, great increase

   (c) shoulder of the *D*–log *H* curve

   ____ .
   slight increase, moderate increase, great increase

2. As development proceeds, how do *slopes* change for exposures lying in the:
   (a) toe of the *D*–log *H* curve

   ____ ;
   slight increase, moderate increase, great increase

   (b) straight line of the *D*–log *H* curve

   ____ .
   slight increase, moderate increase, great increase

3. Gamma is the __________ of the __________ __________ of the *D*–log *H* curve.
   (two words)
4. Gamma has been used as a measure of the contrast associated with __________.
5. "High-contrast" films can produce large values of __________ .
6. Contrast index is a(n) __________ slope.
7. Contrast index is replacing gamma as a measure of the amount of __________.

ANSWERS TO SELF-TEST ON NEGATIVE DEVELOPMENT AND NEGATIVE CONTRAST

In the parentheses following each answer you find the numbers of the items in this chapter that relate to each question and its answer.

1. (a) Slight increase (items 2–4, 7, 8);
   (b) moderate increase (items 11, 12);
   (c) great increase (items 5, 9, 12).
2. (a) Slight increase (items 27, 28);
   (b) great increase (item 15);
   (c) slight increase (items 37–39).
3. Slope, straight line (item 17).
4. Development (item 17).
5. Gamma (item 23).
6. Average (item 47).
7. Development (item 49).

CHAPTER 8 Film Speed—ASA

In practice, a film speed is a number that, when set on a meter dial, along with a meter reading enables the photographer to read out the camera settings (*f*-number and time) that will give, after processing, an excellent negative. The speed value is thus an arbitrary number, having no meaning apart from the photographic system with which it is used.

By the term "system" is meant the entire process, that is: the subject, the lighting, the metering method, the film and its development, and most especially, the definition of "excellence" of the negative. The direct way of finding the speed of a film is almost obvious. Assume a speed value and make a series of exposures bracketing the assumed value; then examine the negatives in the series to find that one having the desired quality with the *least* general exposure level. The decision about negative quality is best based on making prints. Whatever metering method was used in the experiment could be expected to give similar results with a later similar system.

Such an experiment would surely need to be repeated for every different system, since a change in the subject (e.g., from landscape photography to portraiture to line-copy work) would probably demand a change in the definition of "excellence" in the negative. No doubt a significant change in the chemistry of the process would have an effect on the film speed. Thus except as an alert photographer accumulates reliable experience over a long period of time, such a practical method is almost impossibly laborious. In fact, a nearly complete experiment of this kind has been carried out very rarely.

Simply for the sake of efficiency, therefore, it is reasonable to try to find a film speed based on laboratory tests; specifically on data from the sensitometric *D*–log *H* curve. The basic question is, what aspect of the *D*–log *H* curve is related to excellence in the negative? In this section we deal first with the general concept of the speed of a film, and then with the speed method called "ASA."

WHAT YOU WILL LEARN

If you work carefully with this material, you will be able to:

1. Compare the speeds of two films for any specification of image quality;
2. Know the circumstances under which the ASA speed method reliably relates to a practical situation;
3. Find the ASA speed of any film that has been properly tested.

DIRECTIONS. Cover about 2 inches of the right-hand margin of each of the following pages in turn with an opaque sheet of paper. Read carefully the first numbered statement. *Write* the word or phrase that you believe correctly completes the statement. Move the cover sheet down to reveal the correct answer in the margin. Continue in this way until you have finished the section. Work at a rate that is comfortable for you, and no longer than you find pleasant. Begin when you wish.

1. The speed number is based on the exposure needed to produce the desired result. If one film has a speed of 100 and another a speed of 50, the first is said to be ______ (faster, slower) than the second. — faster
2. As the speed number increases, the necessary exposure decreases. Thus to produce the same image the faster film needs ____________ exposure than the slower film. — less
3. The *ratio* of the speeds gives the relationship between the films. In this example, the speed ratio is ____________ to 1. — 2
4. The faster film has twice the speed of the slower film. Put differently, the faster film has one stop more speed than the slower film. The *D*–log *H* curves of the two films might resemble the following sketch.

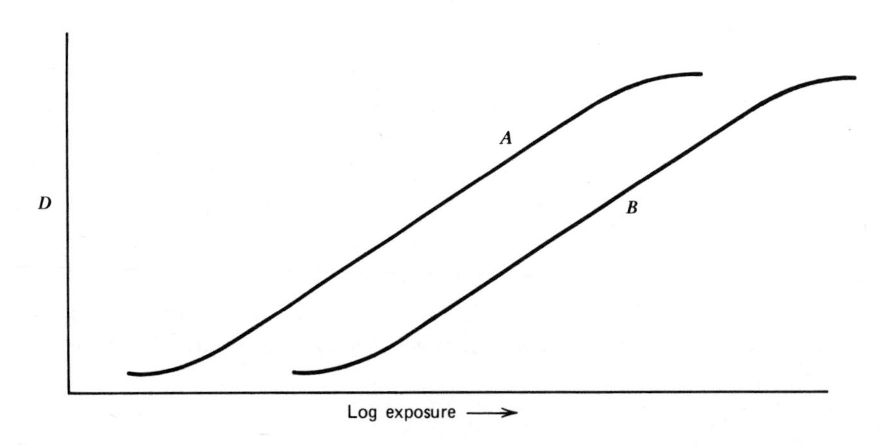

To produce a given density on each of the two curves, the needed exposure is for curve *A* __________ than for curve *B*. — smaller

5. The faster film is represented by curve __________ . — *A*
6. Film *A* is faster because it produces any given density with __________ exposure than film *B*. — less
7. The two curves are everywhere parallel. Thus they have identical slopes and therefore the same __________ . — contrast
8. These two curves would have the same gammas and the same contrast indexes. They would therefore have been developed to the same __________ . — degree, or amount
9. The speed change for these films is shown (in logs) by the *lateral* distance between them, on the log exposure axis, as shown in the sketch repeated as follows:

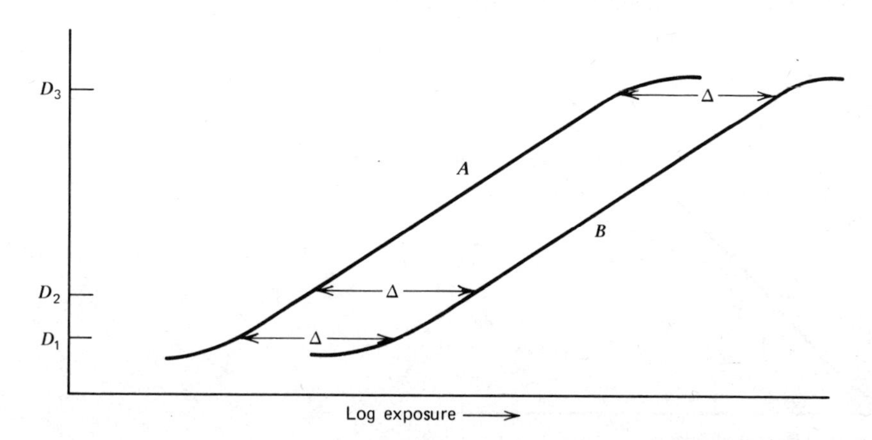

The lateral distance between the curves is indicated by the arrows with the symbol Δ (delta). Since the curves are parallel, the distance Δ is the same at each of the density levels D_1, D_2, and D_3, and would be at the same value for *any* density level.

For a one-stop change in speed, the value Δ would be the log of 2, or __________ . — 0.30

10. Every stop change in speed is shown by a log interval on the log exposure axis of 0.30. If the interval were 0.60, the speed change would be __________ stops. — 2
11. A half-stop speed change would be shown by a value for Δ of __________ . — 0.15
12. For parallel *D*–log *H* curves like the ones above, the lateral interval is the same everywhere, and there is for this case no problem about deciding which film is faster and by how much. A problem *does* arise, however, if the two curves resemble these:

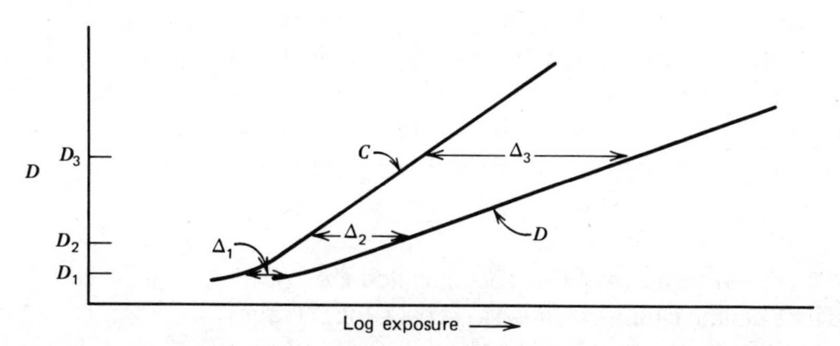

Now, the two curves are *not* parallel, and the value of Δ is __________ at different density levels 1, 2, and 3. — different

13. Because the curve for the film *C* lies everywhere to the left of curve *D*, the faster film is shown by curve ______ .

C

14. Our judgment about the *amount* of speed change, however, will be different, depending on our decision about what density level to choose. At D_1, the value of the lateral distance Δ_1 may be only 0.15. This represents only ______ stop speed change.

½

15. At density level D_2, however, the lateral distance Δ_2 is greater, perhaps 0.30. This represents a full stop change. At D_3 the distance is perhaps 0.60, ______ stops change.

16. Our judgment about the speed comparison will change, for such *nonparallel* curves, dependent on our decision about which density in the negative is critically important. To take an extreme example, consider the curves in the following sketch.

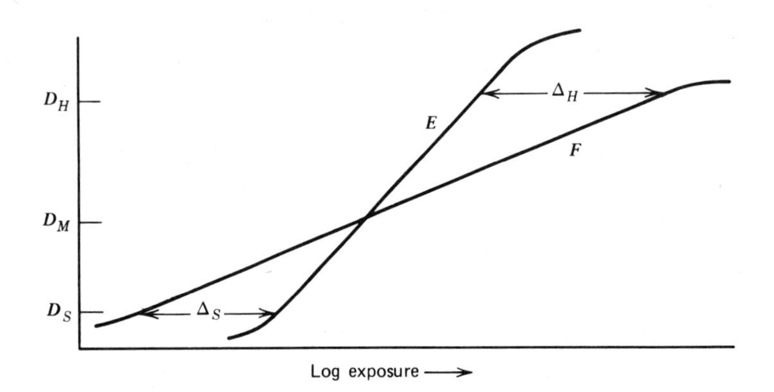

Assume that we based our speed comparison on the midtone density in the negative, and that a desirable midtone density is at D_M. Since the two curves intersect at that density level, they would need the same exposure to produce that density, and we would judge the speed of the two films to be ______ .

the same, or identical

17. Suppose, instead, that we were to judge speed on the reproduction of the shadow, at a density level D_S. We would judge the faster film to be the one lettered ______ .

F

18. Based on the shadow rendition, we would judge *F* to be faster because it requires a ______ exposure to produce the desired shadow density.

smaller

19. On the other hand, if we were to base our decision on the desirable reproduction of the highlights, at a density level D_H, we would judge the faster film to be the one lettered ______ .

E

20. Again, the problem in comparing the speeds of two such unlike films originates in the nonparallelism of the two curves; they were developed to different gammas and different CI values. Because of this difference in the amount of development, we cannot simply use *any* density in the negative as the indication of speed. We *can* use a simple measure of speed only for materials that have been developed to the *same* ______ .

degree, or amount

21. This is the basis for the most common measure of film speed, called the "ASA" speed. To find the ASA speed for a negative film, you need a set of *D*–log *H* curves for different development times, like that in the diagram that follows. You next

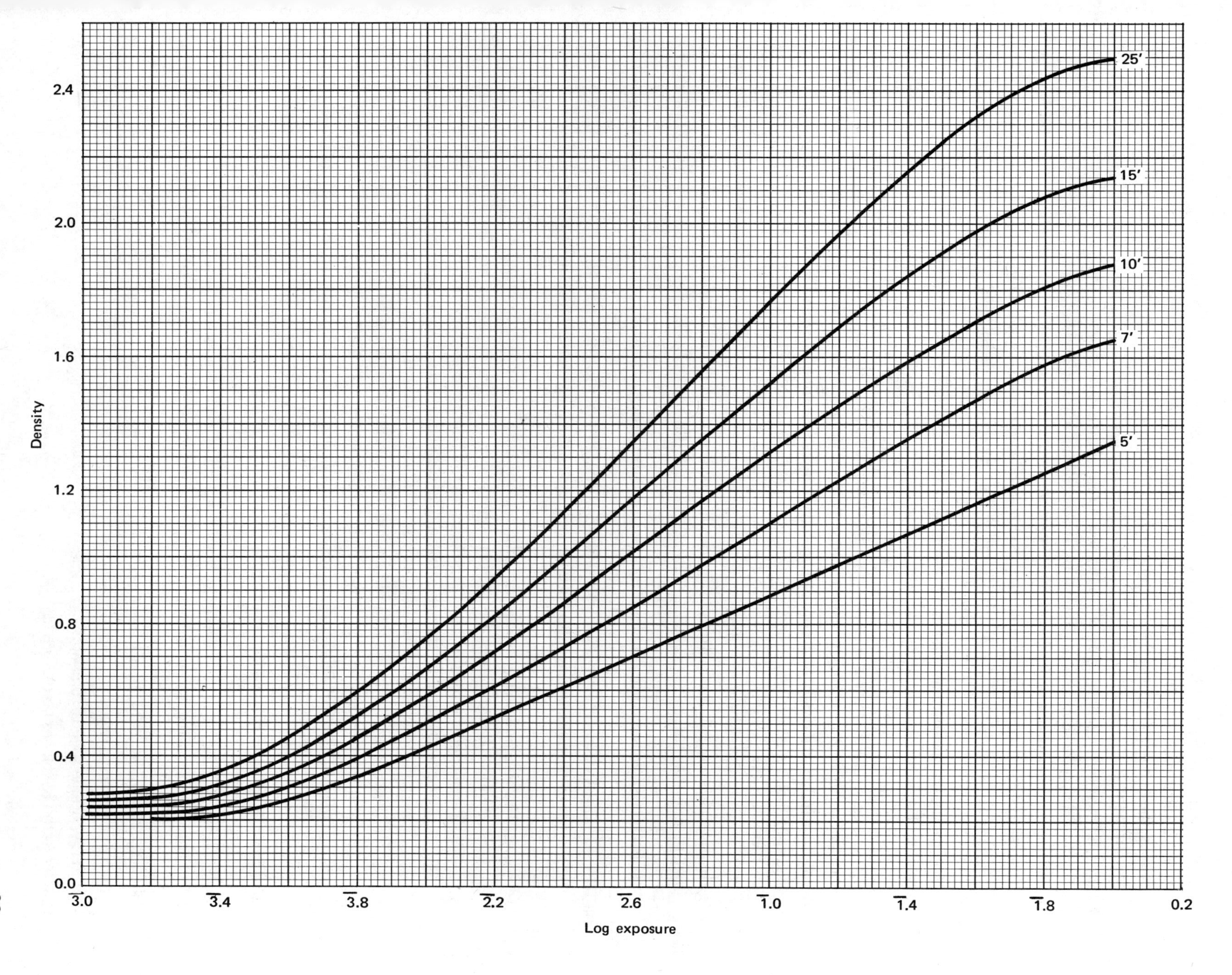
25′
15′
10′
7′
5′
Density
2.4
2.0
1.6
1.2
0.8
0.4
0.0
$\bar{3}.0$
$\bar{3}.4$
$\bar{3}.8$
$\bar{2}.2$
$\bar{2}.6$
$\bar{1}.0$
$\bar{1}.4$
$\bar{1}.8$
0.2
Log exposure

need to find which of the set has a CI close to 0.62. (The method of the published standard differs from this one, but gives nearly the same curve.) For the set that follows, the curve with a CI of 0.62, nearly, is the one for __________ minutes development. — 10

22. Next, on the curve you have selected, find a density equal to 0.10 over base + fog. (The base + fog density is that at which the curve levels out in the toe, in this case __________.) Adding 0.10, you obtain the value __________. — 0.24, 0.34
23. Now read off the curve the log H value that gave a density of 0.34. The log H value is __________. Reading the log H value is difficult because of the small slope of the curve; a difference of 0.02 is trivial. — $\bar{3}.53$
24. To compute the speed value, first find the antilog of the log of the exposure. The antilog of $\bar{3}.53$ is __________. — 0.0034
25. Finally, divide the exposure (0.0034) into 0.8. The answer is the ASA speed. It is __________. (We write the answer to two digits. In use, the nearest value in the standard series of ASA speeds would be chosen; it is 250.) — 240
26. You have followed the steps in the procedure that gives the ASA speed for this film, assuming that it was correctly tested. To summarize, the ASA speed is found from the formula $S_{ASA} = 0.8/H$. H stands for the __________ that gives a density of __________ above base + fog. — exposure, 0.1
27. The formula is correctly applied only to a curve that has been correctly developed, that is, one that has a __________ __________ (two words) close to 0.62. Usually a contrast index within a few hundredths of 0.62 is close enough. — contrast index

NOTES

1. The ASA speed method, strictly defined, demands that the test exposure be with standard daylight. Furthermore, a standard developer must be used.
2. The resulting value can be reliably applied only to general pictorial photography, with typical subject matter.
3. Careful persons use the term *"exposure index"* (EI) for speed values found by methods that differ from the standard, and call the speed "ASA" only when the precise method has been scrupulously followed.
4. The constant 0.8 in the formula for finding the speed merely causes the resulting value to be compatible with existing light-meter dials. Otherwise, it has no significance.

SELF-TEST ON FILM SPEED—ASA

Check your understanding of this chapter by answering the following questions. The correct answers follow the questions.

1. Film X has a speed of 24; film Y has a speed of 100.
   (a) Which film is faster?
   (b) How do the speeds compare?
2. Refer to the D–log H curves sketched.

Self-Test on Film Speed—ASA

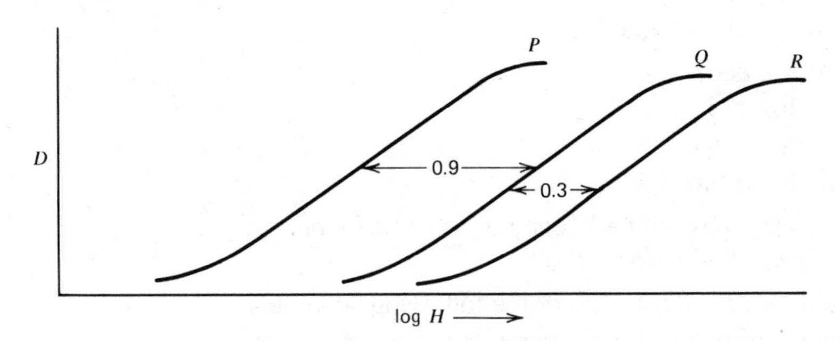

(a) Which film, P, Q, or R, is fastest?

(b) Which film is slowest?

(c) By how many stops do the speeds of Q and R compare?

(d) By how many stops do the speeds of P and Q compare?

3. Refer to the D–log H curves sketched in the following diagram.

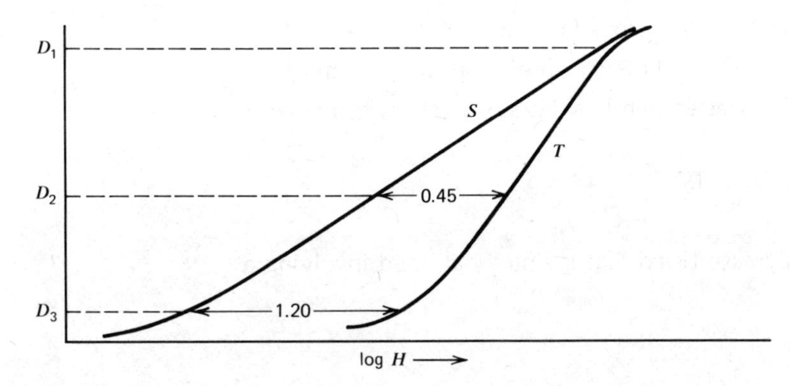

(a) Judging by the exposure needed to produce a high density (D_1), how do the speeds of the two films compare?

(b) Judging by the exposures needed to produce a midtone (D_2), how do the speeds compare (i.e., which is faster, and by how much)?

(c) Judging by the exposures needed to produce a low density (D_3), how do the speeds compare?

4. Refer to the D–log H curves sketched as follows.

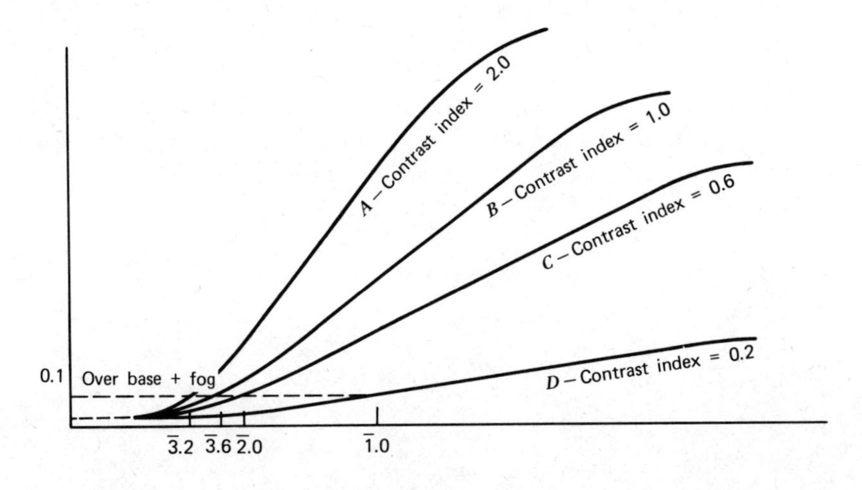

(a) For which curve can the ASA speed be found?
(b) For that curve, what is the ASA speed:
  (i) log H at 0.1 over base + fog = ?
  (ii) the antilog of that log H = ?
  (iii) the answer to (2) divided into 0.8 = ?

5. For which of the following light sources may the ASA speed of a film be found: (a) tungsten; (b) fluorescent; (c) daylight; (d) electronic flash.
6. The published ASA speed of a film is 125. For which of the following situations would you expect it to be a reliable guide: (a) landscape photography; (b) studio portraiture; (c) a sunset scene; (d) the interior of a supermarket.

ANSWERS TO SELF-TEST FILM SPEED—ASA

1. (a) Y; (b) Y is four times as fast as X, or two stops faster than X.
2. (a) P; (b) R; (c) Q is one stop faster than R; (d) P is three stops faster than Q.
3. (a) They are practically equal; (b) S is faster than T by 1½ stops; (c) S is faster than T by 3 stops.
4. (a) Curve C (bi) $\overline{3}.6$; (bii) 0.004; (biii) 200.
5. Daylight.
6. Landscape photography, the only case listed that is "pictorial" and involving a subject of normal range.

CHAPTER 9
Film Speed—Exposure Index

There are many situations in which we wish to find a film speed but cannot match the ASA requirements. The most obvious case is one in which we cannot (or do not wish to) match the required contrast index of 0.62. It is common in the zone system, for example, to alter development time for negatives to adjust them for varying subject contrasts. It is then necessary to resort to a different method of finding an exposure index (EI).

In this chapter we first review and extend some of the concepts dealt with in Chapter 6. There we defined the useful range of exposures for "normal" development, that is, for a contrast index of nearly 0.6.

WHAT YOU WILL LEARN

If you work carefully with the material in this chapter, you will learn:

1. That when camera exposure is reduced, it is the dark tones of the subject (the shadows) that suffer loss.
2. That the loss of quality in the shadows is caused by the reduced slope (eventually to zero) in the toe of the *D*–log *H* curve.
3. That for excellent negatives of pictorial subjects the least slope must be at least 0.3 times the contrast index.
4. That the preceding statement justifies a method of

speed determination that can apply to a *D*–log *H* curve developed to any contrast index.

5. How to find an exposure index based on the rule stated in 3 above.

DIRECTIONS. Cover about 2 inches of the right-hand margin of each of the following pages in turn with an opaque sheet of paper. Read carefully the first numbered statement. *Write* the word or phrase that you believe correctly completes the statement. Move the cover sheet down to reveal the correct answer in the margin. Continue in this way until you have finished the section. Work at a rate that is comfortable for you, and no longer than you find pleasant. Begin when you wish.

1. Consider a film developed so as to produce a *D*–log *H* curve as sketched in the following diagram. The film has been developed to a small CI, about 0.45. Shown in the sketch is an exposure situation, where the range of subject exposures is 2.0 in logs, and with all the exposures falling on the straight line of the *D*–log *H* curve.

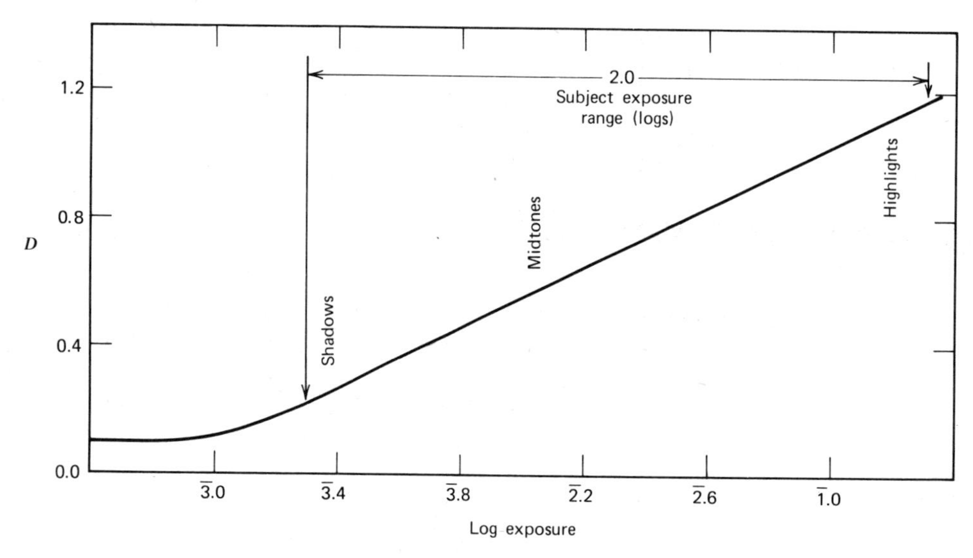

All parts of the subject—shadows, midtones, and highlights—were exposed on the straight line of the *D*–log *H* curve. Since the straight line is the region of uniform, constant slope, all parts of the negative image would have the same ___________ .

contrast

2. Suppose now a second negative made of the same subject but with reduced camera exposure. All the subject exposures would move to the ___________ .

left

3. If the reduction in camera exposure were 1 stop, the subject exposures would move to the left a distance of ___________ in logs.

0.3

4. The new situation is as shown on page 105, where the dotted arrows represent the extreme subject exposures in the first case and the heavy arrows represent the extreme exposures in the second case. The highlight and midtone subject exposures still fall on the ___________ ___________ .
(two words)

straight line

5. Therefore, in the second case the highlights and midtones would be reproduced with the same ___________ as in the first case.

contrast

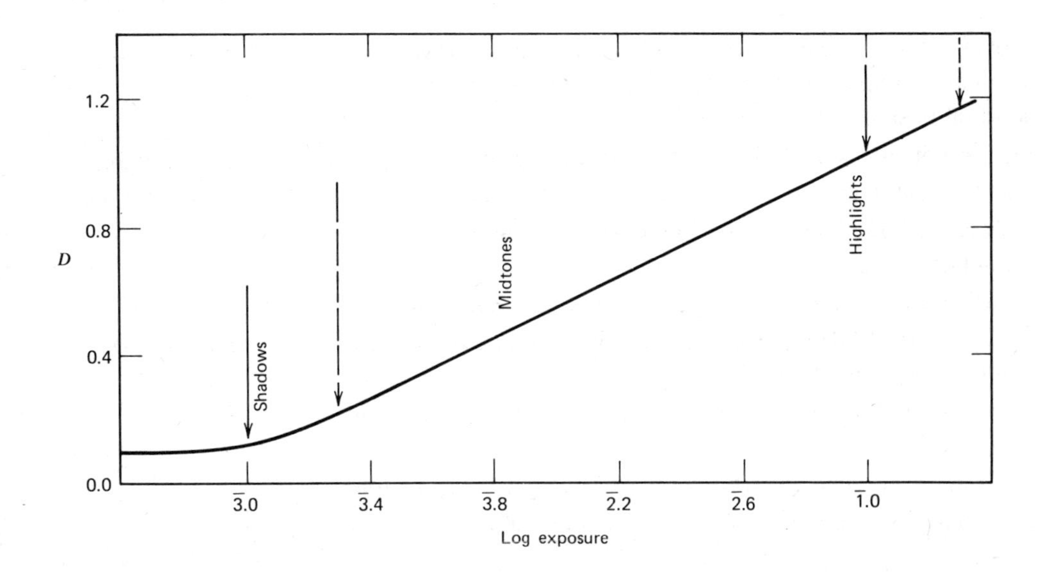

6. Even with a reduced exposure level of one stop, the highlights and midtones would have the same contrast and the same detail. In the second case, however, the shadows have moved into the ____________ of the curve. — toe
7. In the toe of the curve, the slope is ____________ than in the straight line. — less
8. Since the shadows in the second case fall in the toe, the region of lesser slope than in the straight line, the shadow contrast would become ____________ . — less
9. Consider yet a third case, with one stop less exposure again, as shown in the following sketch. The subject exposures have been moved again to the left, an

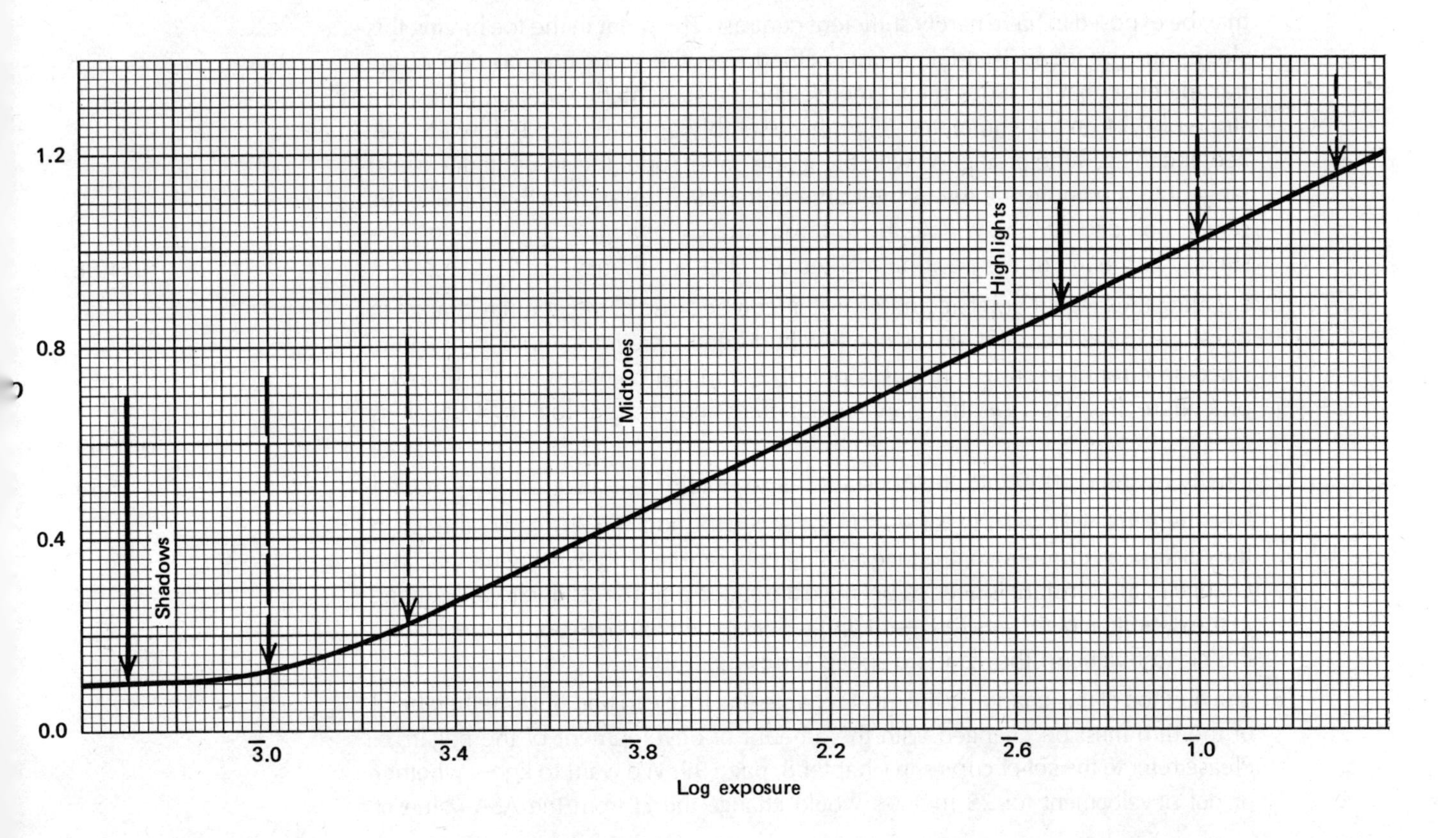

additional distance of 0.3; the extreme exposures lie at the heavy arrows. The dotted arrows are the extreme exposure positions in the first case, and the dashed arrows the corresponding positions in the second.

Even in the third negative, with two stops less camera exposure than in the first, the negative contrast is still the same for the ______ and the ______ .

midtones, highlights (either order)

10. Now, however, the shadows have been moved to the extreme toe of the curve, where the slope is practically ______ .

zero

11. Since the slope is nearly zero for the shadows, they would have in the negative almost no ______ . In the third negative, the shadows have so little contrast that they would certainly not be printable.

detail, or contrast

12. The preceding analysis shows that as camera exposure is reduced, it is the ______ tones that suffer.

shadow

13. For this reason, it is sensible to base the exposure index on the right reproduction of the shadows, and specifically on the *slope* needed to produce sufficient shadow ______ .

detail, or contrast

14. Experiments with pictorial photographic images carried out many years ago showed that for excellent negatives the shadow slope must be at least 0.3 of the average slope which we measure by CI. The 0.3 factor was found to be correct for *any* CI. We can use this rule, therefore, for curves representing development either more or less than that demanded by the ASA method. The curve used for illustration in this section was not developed to the CI of 0.62 required to find the ASA speed. Its CI is considerably less than 0.62, in fact about 0.46. To find the EI for this material so developed, first multiply the CI by 0.3. The result is ______ .

0.14

15. The value 0.14 is the minimum slope, in the toe of the curve, at which the shadows may be exposed to have barely sufficient contrast. The point in the toe having this slope must be found by trial. A fast method uses the gammeter portion of the transparency that includes the CI meter. Place a straightedge so that it lies at a slope of 0.14 on the gammeter. Slide the assembly upward from below the toe, keeping the gammeter aligned with the graph paper, until the straightedge is just tangent to a point in the toe of the curve in item 9. The result will look like the sketch on page 107. Now note the log exposure value at the tangent point. It is slightly to the left of the position marked $\overline{3}.0$, that is, about ______ .

$\overline{4}.96$

16. Locating the tangent point accurately is difficult because of the gradual change in the slope in the toe. Differences of a few hundredths are immaterial in practice, since an error of even 0.1 in logs is only 1/3 stop. To compute the EI, follow a procedure like that for the ASA speed: first find the antilog of $\overline{4}.96$; it is ______ .

0.0009

17. Finally divide 0.0009 into the constant *0.5* to obtain ______ .

560

Note here that we make the constant in the speed formula 0.5 (instead of the 0.8 in the ASA speed formula) to produce a value compatible with existing meters. The change in the constant is needed because the speed point shifts due to the change in the definition of the desired result. In practice, you would use as a meter setting the closest value in the standard speed series (i.e., 500).

18. The method of this chapter is the most reliable one to use to find out whether the EI of the film must be changed with the amount of development of the negative. Please refer to the set of curves in Chapter 8, page 99. We want to know whether or not development for 25 minutes would change the EI from the ASA value of

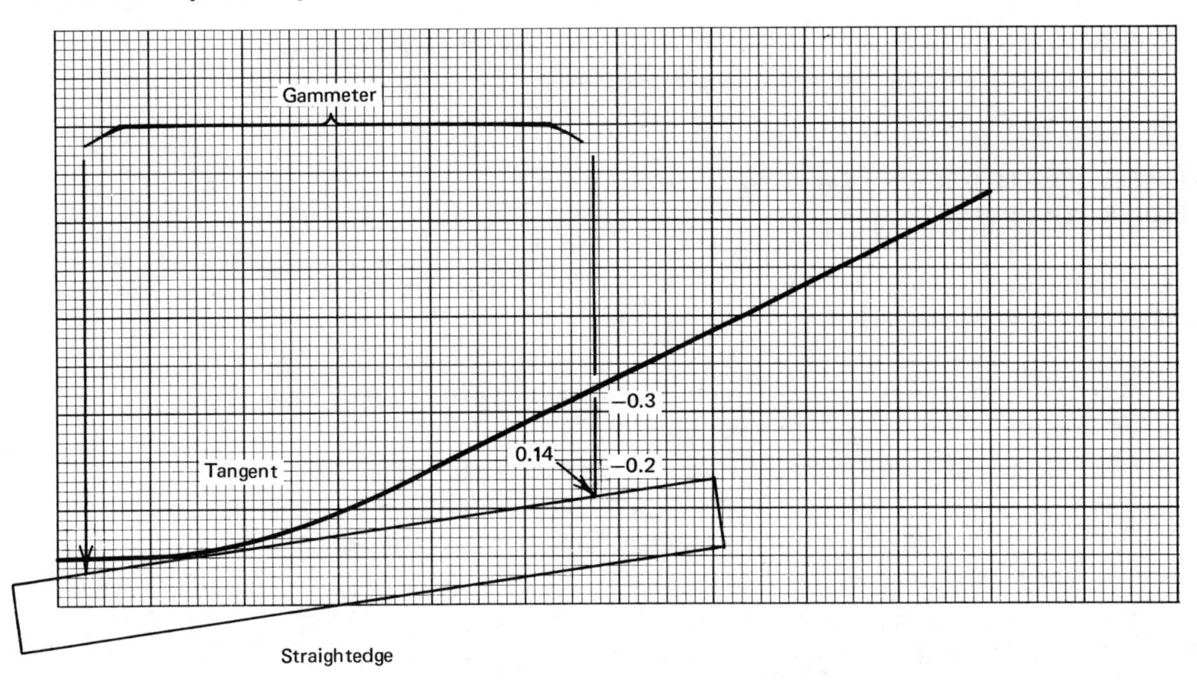

240 found for the 10-minute curve. First find the CI. It is about ________ . — 0.92

19. Now you need to find the point in the toe of the curve where the slope is 0.3 times 0.92, or ________ . — 0.28
20. Following the method of items 15–16 above, the log exposure value is about ________ . — $\bar{3}.35$
21. The antilog of $\bar{3}.35$ is ________ . — 0.0022
22. The EI is found from 0.5/0.0022, and is ________ . — 230

    Comparing this value with the ASA speed you found before, the conclusion is that the increase of development time has certainly *not* increased the EI.

SELF-TEST ON FILM SPEED—EXPOSURE INDEX

Check your understanding of this chapter by answering the following questions. The correct answers follow.

1. (a) As camera exposure is reduced, in what direction do the subject exposures move (relative to the *D*–log *H* curve)?
   (b) As camera exposure is increased, in what direction do the subject exposures move (relative to the *D*–log *H* curve)?
   (c) (Trick question.) As camera exposure is reduced, in what direction does the *D*–log *H* curve move?

2. As camera exposure is reduced from a correct exposure level, determine which tones of the subject: (a) lose in quality (b) remain relatively unchanged in quality.

3. What characteristic of the *D*–log *H* curve is directly related to negative quality?

4. Apply the method of the text to find an EI for the 5-minute curve of the set on page 99. The CI of the 5-minute curve is approximately 0.44.

ANSWERS TO SELF-TEST ON FILM SPEED—EXPOSURE INDEX

In the parentheses after each answer you will find the items in this chapter that relate to the question and its answer.

1 (a) When camera exposure is reduced, subject exposures move to the *left* on the log-exposure axis of the *D*–log *H* curve (items 2, 9).
  (b) When camera exposure is increased, subject exposures move to the *right* of the log-exposure axis of the *D*–log *H* curve (converse of items 2, 9).
  (c) The *D*–log *H* curve *never* moves as a result of changed exposure level.
2. As camera exposure is reduced from a correct level
  (a) The shadow (dark) subject tones lose quality (items 8, 10–12).
  (b) The midtone and highlight areas remain relatively unchanged (items 4–6).
3. Negative quality is associated with the slope in the toe of the *D*–log *H* curve, which for pictorial photography should be at least 0.3 of the average slope (items 13 and 14).
4. Contrast index of $0.44 \times 0.3 = 0.13$ approximately. Using the method of items 15–16, the log exposure at the tangent point is about $\overline{3}.35$. The antilog of $\overline{3}.35$ is 0.0022. From the formula $0.5/H$, the exposure index is approximately 230.

CHAPTER 10

Photographic Papers—Print Tones

Most photographic images are presented to the observer in the form of a print on photographic paper. The characteristics of photographic papers are exceptionally important, since they significantly affect the quality of the image and often limit the success of the entire photographic process.

To test photographic papers, a sensitometer may be used. It is often simpler and more practical to use as a test object a transparent tablet in a contact or projection printer. The densities of the original step tablet control the illuminances on the paper sample in corresponding areas and (with the exposure time included) determine the exposures the sample receives. With this section of text you have samples of photographic papers exposed in this way. (Samples are in an envelope at the back of the book.)

The processed sample is measured with a reflection densitometer. This instrument produces on the sample a fixed illuminance; the light reflected from the sample is received by a photocell, and a scale permits reading of the density of any given area directly. Schematically, the densitometer is as in the sketch on page 110.

Reflection density, like transmission density, is defined as the logarithm of the ratio of the initial to the reflected light. $D_r = \log(E_o/E_r)$, where E_o is the illuminance at the sample and E_r is the reflected light received by the photocell. Alternatively, $D_r = \log(1/R)$, where R is the reflectance of the sample, the ratio of E_r to E_o. If the sample reflects half the light it receives, R is 0.5, $1/R$ is 2.0, and the density is the log of 2.0, or 0.3. See Chapter 2, items 34–50.

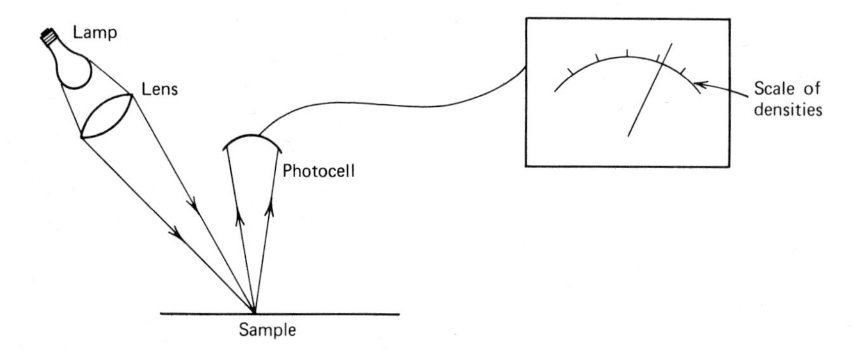

In Chapters 4 and 5 you found out that for negative materials there are two ranges of interest:

1. The range of densities that the photographic material can produce;
2. The range of log exposures over which the material produces useful images.

These same ranges are important for photographic papers.

In this chapter we consider the factors that determine the tonal range that can be obtained from a photographic paper.

WHAT YOU WILL LEARN

If you work carefully with the following material, you will:

1. Know the characteristic of a paper that produces brilliant highlights;
2. Know the characteristic of the paper that produces deep black tones;
3. Be able to find the range of tones that a paper can produce.

DIRECTIONS. Cover about 2 inches of the right-hand margin of each of the following pages in turn with an opaque sheet of paper. Read carefully the first numbered statement. *Write* the word or phrase that you believe correctly completes the statement. Move the cover sheet down to reveal the correct answer in the margin. Continue in this way until you have finished the section. Work at a rate that is comfortable for you, and no longer than you find pleasant. Begin when you wish.

DENSITY RANGE OF PAPERS

1. The range of densities, and thus of tones, that a paper *can produce* is determined completely by just two factors. One of these is how near to pure white is the paper base, or the paper stock. The density of the base determines the lightest possible tone, and thus the reproduction of the ____________ of the subject. — highlights
2. One of the samples with this text was coated on the "whitest" possible base. It has a reflectance of about 0.85. Its density is then log(1/0.85) or log 1.17, which is about ____________ . — 0.07

3. Another of the samples was coated on an "ivory" stock. The reflectance of this base is about 0.80, and the base density is thus about __________ . — 0.10

4. The measured density of the white stock is only a little less than that of the ivory stock—0.07, compared with 0.10. When you place the two samples side by side, you can easily see that the lightest tone is certainly darker for the __________ stock. — ivory

5. This comparison shows that your eyes are very sensitive to slight differences in the light tones of the print material. If you want the lightest possible (most brilliant) highlights in a print, you will use a paper coated on a __________ stock. — white

   Although comparison easily shows the difference in the lightest tones for different paper stocks, when each image is seen separately, we tend to accept as "white" the lightest tone the paper can give. For this reason, it is common to adjust a reflection densitometer so that it reads zero when placed over an unexposed, fixed-out part of the test sample.

6. The second factor that determines the range of tones obtainable in a print on photographic paper is the darkest "black," measured by the maximum density (D_{max}) it can produce. D_{max} is entirely determined by the paper *surface* and is almost unaffected by paper grade or developer. The greatest value of D_{max} is found in a paper having the glossiest possible surface. If the reflectance of the darkest tone is 0.01, the density (from $D_r = \log(1/R)$) is __________ . — 2.0

7. D_{max} for the glossy sample you have is even greater, approximately 2.2. On the other hand, a dull-surface ("mat") paper will have a reflectance for the darkest possible tone of as much as 0.03, for a D_{max} of only about __________ . — 1.5

8. A comparison of the mat and glossy samples shows that the "black" is much darker for the __________ sample. — glossy

9. Your samples show approximately the extreme differences obtainable from papers of different surfaces. If a paper has a surface between glossy and mat, like semigloss or "luster," it will have a D_{max} between 1.5 and 2.2. If you want a print with the deepest possible black tone, you will print on a paper having a __________ surface. — glossy

10. Because ferrotyping a print (drying it in contact with a shiny metal surface) makes the surface glossier, it makes the dark tones of the print __________ . — darker

11. A mat print before drying is covered with a thin film of water. Since this shiny film of water disappears as it dries, after drying the "blacks" of a mat print become __________ . — lighter, less dark (etc.)

12. If a mat print is waxed or sprayed with a varnish, the dark tones become __________ . — darker

13. To summarize, the lightness of the lightest tones in a print is determined by the paper __________ ; the darkness of the darkest tones is determined by the paper __________ . — base or stock, surface

14. The greatest range of tones you can obtain in a print is measured as the difference in the extreme densities. For a glossy print on a white stock, the range is from about 0.07 to 2.2, a difference of __________ . — 2.13

15. If we round off this difference to 2.1 and take its antilog, we have the greatest possible ratio of luminances in a print on such a paper. The antilog is ________ . — 130

16. Thus the greatest possible luminance ratio for a glossy print on a white stock is 130 to

1. Recall that the average outdoor scene includes about 7 zones, each zone being 0.3 in logs, for a total log range in the scene of ______ . — 2.1

17. Such a scene covers about 7 scale intervals on modern light meters. Note that there is a near match between the range of such a scene and the maximum range possible with a *glossy* paper. On the other hand, a print on a mat paper has a range from 0.10 to only about 1.70, for a difference of ______ . — 1.60

18. Taking the antilog of this difference (1.60) you find the value ______ . — 40

19. Thus the greatest possible tonal range in a mat print is about ______ to 1. — 40

20. The range of tones in the typical outdoor subject is such that it cannot possibly be reproduced in a print made on a ______ surface. — mat

21. To reproduce such a range, you must make a print on a ______ surface. — glossy

22. Since a mat paper has a density range of only about 1.4, it can reproduce the range of a subject covering slightly less than ______ zones. — 5

23. We can apply the concepts developed above to photographs reproduced in a newspaper. Newspaper stock is far from "white," and because of the extremely rough surface, the darkest tone is far from "black." The greatest possible range of densities in ordinary newspapers is usually less than 1.3. The antilog is ______ . — 20

24. Thus the maximum tonal range in newspaper reproduction is only about 20 to 1; this is why the images are so poor, especially by comparison with a glossy print. To make newspaper reproduction better, better paper would have to be used—a stock more nearly white, and a ______ surface. — smooth, or glossy

25. Magazine reproduction of photographic images is usually better than in newspapers because magazines use paper that is ______ and ______ . — white, glossy (either order)

26. Some persons speak of the "number of tones" in a print. The number of tones is not really very important. It is, however, determined by two factors: (a) the range of tones and (b) the least difference in tone that the viewer can distinguish. The least perceptible difference varies with the density level and also with the tone of nearby areas. A reasonable average value is 0.04. If the total density range is 2.2 and the least perceptible difference is 0.04, the quotient of these numbers is the number of perceptible tones; it is ______ . — 55

27. Thus there are about 50 or more perceptibly different tones in a glossy print with a full range from "white" to "black." If the print is mat, and has a density range of 1.4, dividing by 0.04 gives ______ . — 35

Thus there are about 35 perceptibly different tones, even in a mat print having a small density range. Even if the preceding calculations are subject to question in detail, there are certainly several dozen tones that exist, in this sense, in any full-range print. You may compare this statement with the often-heard remark that there are only 5, or 7, or some such small number of tones in a print.

SELF-TEST ON PHOTOGRAPHIC PAPERS—PRINT TONES

Check your understanding of this chapter by answering the following questions. The correct answers follow.

1. The lightest possible highlights can be produced in a print made with a ______ stock.

2. The darkest possible tones can be produced in a print having a __________ surface.
3. Drying a print with the surface in contact with a very smooth metal surface is called "ferrotyping." The result is a very glossy surface on the print. The effect of this process is to make ____

   dark tones darker, light tones lighter, no change in the tones.
4. A semigloss paper gives a minimum density of 0.10 and a maximum density of 1.60. The tonal range in logs is __________ .
5. The paper in question 4 could reproduce in the print a subject having a range of about __________ zones.
6. Newspaper photographic reproductions on recycled newsprint can produce at most a tonal range in logs of 1.2. For this case, the subject should have a range of no more than __________ zones.

ANSWERS TO SELF-TEST ON PHOTOGRAPHIC PAPERS—PRINT TONES

In the parentheses after each answer you find the numbers of the items in this chapter that relate to each question and its answer.

1. White (items 1–5).
2. Glossy (items 6–8).
3. Dark tones darker (item 10).
4. 1.50 (items 14, 15).
5. 5 (items 16–17).
6. 4 (items 16–17, 23–24).

CHAPTER 11

Photographic Papers—*D*–log *H* Curves and Scale Index

In Chapter 10 you saw that there are two ranges that are of importance for photographic papers and that the *tonal* range is determined by the physical characteristics of the paper itself. The other range is that of the *log exposures* to which the paper can respond; this is the subject of this chapter.

WHAT YOU WILL LEARN

If you work carefully with the following material, you will be able to:

1. Plot *D*–log *H* curves for photographic papers from a test using the printer as a substitute for a sensitometer;
2. From plotted curves, find the *useful* range of densities and log exposures;
3. Choose the right paper for any reasonable negative, when the data are available.

DIRECTIONS. Cover about 2 inches of the right-hand margin of each of the following pages in turn with an opaque sheet of paper. Read carefully the first numbered statement. *Write* the word or phrase that you believe correctly completes the statement. Move the cover sheet down to reveal the correct answer in the margin. Continue in this way until you have finished the section. Work at a rate that is comfortable for you, and no longer than you find pleasant. Begin when you wish.

The samples of photographic papers you have been using were exposed to a negative step tablet by contact, using the light from a projection printer as a source. This method is similar to the use of a sensitometer but better because it is closer to practice.

The illuminance on the exposure plane was 63 lux (meter-candles). The exposure time was 5 seconds.

1. Without the step tablet the exposure (from the $H = Et$ formula) would have been ________ lux-seconds (meter-candle-seconds).

   315

2. Without the step tablet, the log exposure would have been ________ .

   2.50

3. To find the log exposure received by any part of the paper sample, you need to subtract the density of the corresponding part of the negative step tablet from the log-exposure value found in item 2 above. Thus if one step of the tablet had a density of 0.30, subtract 0.30 from 2.50 to get ________ .

   2.20

4. The value 2.20 is the log exposure received by the paper sample under that patch of the negative step tablet that had a density of 0.30. The greatest density of the negative step tablet was 3.00. Subtracting that density from the log *H* value found in item 1 above, you find the log *H* value of ________ . (See Appendix Section A5 for the subtraction of logs.)

   $\bar{1}.50$

5. This log *H* ($\bar{1}.5$) is the *smallest* log exposure that the paper sample received, because it came from the densest part of the negative step tablet that absorbed the greatest fraction of the initial light. The smallest density of the negative step tablet was practically 0.00. Under this patch of the step tablet the sample received the greatest exposure, a log *H* value of ________ .

   2.50

6. The densities of the negative step tablet increased uniformly from patch to patch, by a value of 0.15. Thus beginning from 0.00, the negative density values were: 0.00; 0.15; ________ ; ________ . . .

   0.30; 0.45 . . .

7. As the densities of the test step tablet increase, the log *H* values received by the sample ________ .

   decrease

8. Every change in the density of the negative step tablet produces the same change in the log exposure received by the sample. Therefore we can mark the log *H* axis of the *D*–log *H* curve for the paper as on the next page; there are 21 tick marks on the log *H* axis, one for each density of the negative step tablet and for each corresponding log *H* value. The negative step-tablet densities are reversed from their normal plotting order because small negative densities produce ________ exposures.

   large

9. Two blanks have been left in the series of negative step-tablet densities; in order from left to right the correct values are ________ and ________ .

   2.25, 0.45

10. Two blanks have also been left in the series of log *H* values; in order from left to right these values are ________ and ________ .

    0.70, 1.60

11. The reflection densities for the glossy paper sample follow; they are referred to the density of the stock as zero, and in order of increasing exposure: 0.00, 0.00, 0.00, 0.00, 0.00, 0.01, 0.03, 0.05, 0.22, 0.56, 1.15, 1.75, 1.97, 1.99, 2.00, 2.00, 2.00, 2.00, 2.00, 2.00, 2.00. Please plot the preceding numbers opposite the corresponding log *H* values and draw in a smooth curve. You have plotted the *D*–log *H* curve for this paper. You may compare your plot with the correct graph that follows.

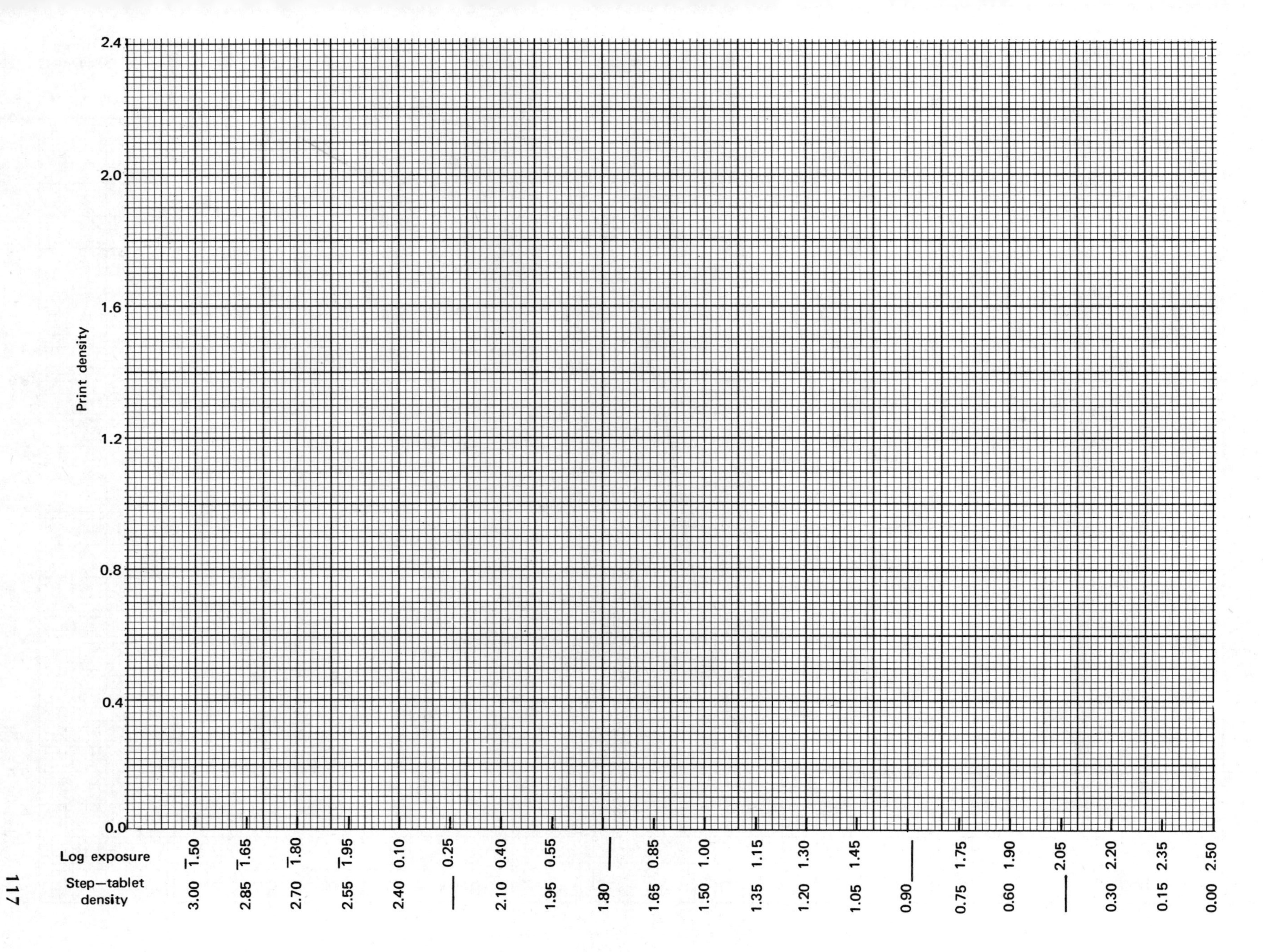
2.4
2.0
1.6
1.2
0.8
0.4
0.0
Print density
Log exposure
Step–tablet density
$\bar{1}.50$ 3.00
$\bar{1}.65$ 2.85
$\bar{1}.80$ 2.70
$\bar{1}.95$ 2.55
0.10 2.40
0.25
0.40 2.10
0.55 1.95
1.80
0.85 1.65
1.00 1.50
1.15 1.35
1.30 1.20
1.45 1.05
0.90
1.75 0.75
1.90 0.60
2.05
2.20 0.30
2.35 0.15
2.50 0.00

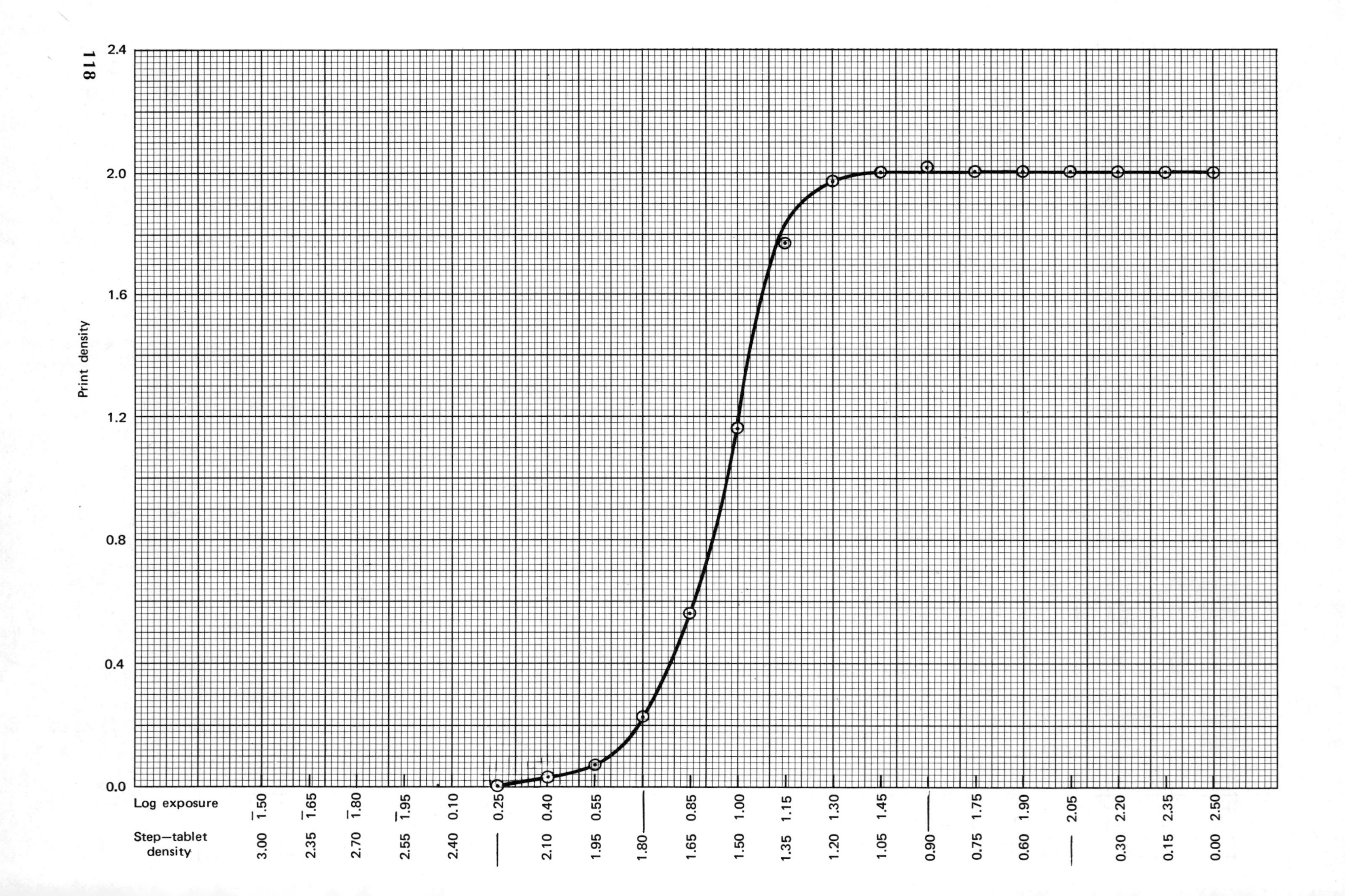

Print density
2.4
2.0
1.6
1.2
0.8
0.4
0.0
Log exposure
Step—tablet density
$\bar{1}$.50 $\bar{1}$.65 $\bar{1}$.80 $\bar{1}$.95 0.10 0.25 0.40 0.55 0.85 1.00 1.15 1.30 1.45 1.75 1.90 2.05 2.20 2.35 2.50
3.00 2.35 2.70 2.55 2.40 2.10 1.95 1.80 1.65 1.50 1.35 1.20 1.05 0.90 0.75 0.60 0.30 0.15 0.00

Note that this curve has a continually changing slope. It is typical of papers, which, unlike negative materials, have almost no region of constant slope, that is, no ________ ________ . (two words) — straight line

12. No density was produced in the sample for log-exposure values less than about ________ . — 0.25
13. Also, no density was produced in the paper sample for negative step-tablet densities greater than about ________ . — 2.25
14. The slope of the paper curve continually *increases* between log H values of ________ and ________ . — 0.25 and 1.00
15. The slope of the curve continually *decreases* between log H values of ________ and ________ . — 1.00 and 1.45
16. The maximum density in the print is reached at a log H value of about ________ . — 1.45
17. All log H values greater than 1.45 produce the same maximum density in the print; for such values the slope of the curve is practically ________ . — zero
18. Maximum density in the print is produced for all negative step-tablet densities ________ than about 1.05. — less
19. In the language of the photographer, "blocked highlights," that is, blank undifferentiated white tones, are produced for all negative densities more than about ________ . — 2.25
20. Using similar language, no shadow detail would be produced in the print for any negative densities less than about ________ . — 1.05
21. Differences in print density (and thus print tones) would be produced only for negative densities lying between ________ and ________ . — 1.05, 2.25
22. We know the *useful* limits of the paper curve for general pictorial photography from experiments in which observers were asked to select the prints they preferred when a negative of a given scene was used to make many different prints. A preferred print has the lightest tone at a density of about 0.04. For the sample curve you have plotted, a density of 0.04 is produced at a log-exposure value of ________ . — 0.45
23. You have found the *minimum useful* log exposure for this paper to be ________ . — 0.45

    The *maximum* useful point from the same kind of experiment was found to be at a density that is 90% of the greatest density the paper can produce. The greatest density for the glossy sample was ________ , and 90% of that density is ________ . — 2.00, 1.80
24. From the curve you have plotted, 90% of the maximum density (1.80) was produced by a log-exposure value of ________ . — 1.17
25. You have found the *maximum useful* log exposure for this paper to be ________ . — 1.17

    The useful log-exposure *range* is the difference in the minimum useful log exposure (answer to item 22) and the maximum useful log exposure (answer to item 24). The useful exposure range is ________ . — 0.72
26. The useful exposure range is also called the "scale index" of the paper. You should agree that the useful exposure range (scale index) is also the difference in the densities of the negative that will produce all of the tones in the print between the minimum and maximum useful points. Thus the scale index of this paper sample is ________ , and to print well on this paper a negative should have a density difference of ________ . — 0.72, 0.72

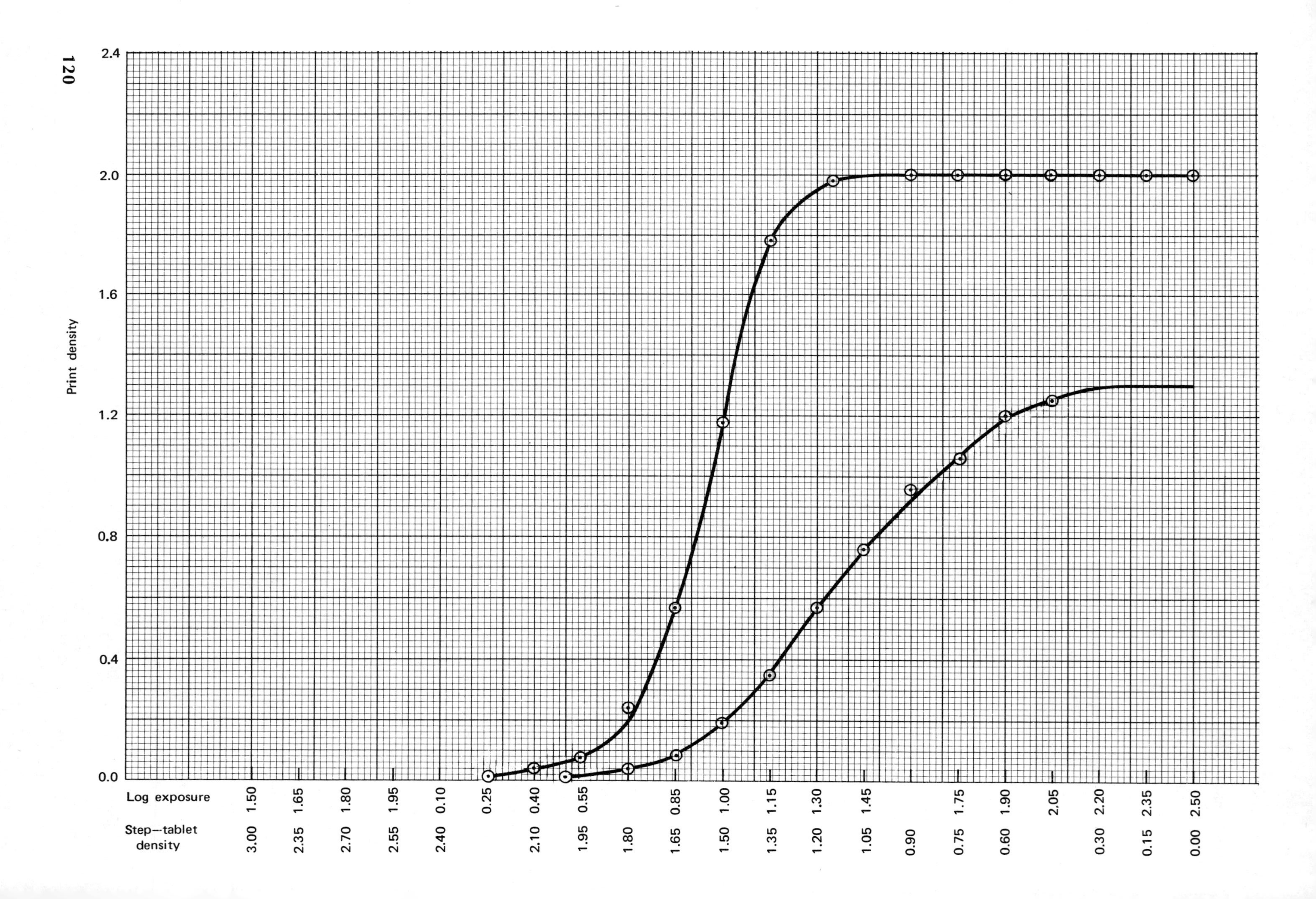
Print density
2.4
2.0
1.6
1.2
0.8
0.4
0.0
Log exposure
1.50 1.65 1.80 1.95 0.10 0.25 0.40 0.55 0.85 1.00 1.15 1.30 1.45 1.75 1.90 2.05 2.20 2.35 2.50
Step–tablet density
3.00 2.35 2.70 2.55 2.40 2.10 1.95 1.80 1.65 1.50 1.35 1.20 1.05 0.90 0.75 0.60 0.30 0.15 0.00

27. If you know the scale index of the paper, you also know the necessary difference in the densities of the ____________ that will print well on the paper. — negative

28. This principle permits you to match the negative to the print material. To apply this rule to the other paper sample, please plot the following densities for that paper on the same graph paper you used for item 11 above. The exposure conditions were the same, and thus these new data should be plotted against the same exposure values as before. 0.00, 0.00, 0.00, 0.00, 0.00, 0.00, 0.00, 0.01, 0.04, 0.09, 0.19, 0.35, 0.56, 0.76, 0.95, 1.09, 1.20, 1.25, 1.29, 1.30, 1.30. In the graph on page 120 you find curves for both samples correctly plotted. — **ins. W 93**

29. Now find the scale index for this paper. First find the log-exposure value for a paper density of 0.04. It is ____________ . — 0.70

30. For this sample the maximum density is 1.30, and 90% of 1.30 is ____________ . — 1.17

31. The log *H* value that produced this maximum useful density (1.17) is ____________ . — 1.85

32. Find the difference between the answers you have found in items 29 and 31; it is ____________ . — 1.15

33. The scale index for this mat paper is ____________ . — 1.15

34. The difference you would need between the extreme densities in a negative that would produce a full range of tones on this mat paper is ____________ . — 1.15

35. The scale index of a photographic paper is roughly related to the paper *grade*. Papers called "soft" are graded 0 and 1; they have large-scale index values, as sketched in the following diagrams.

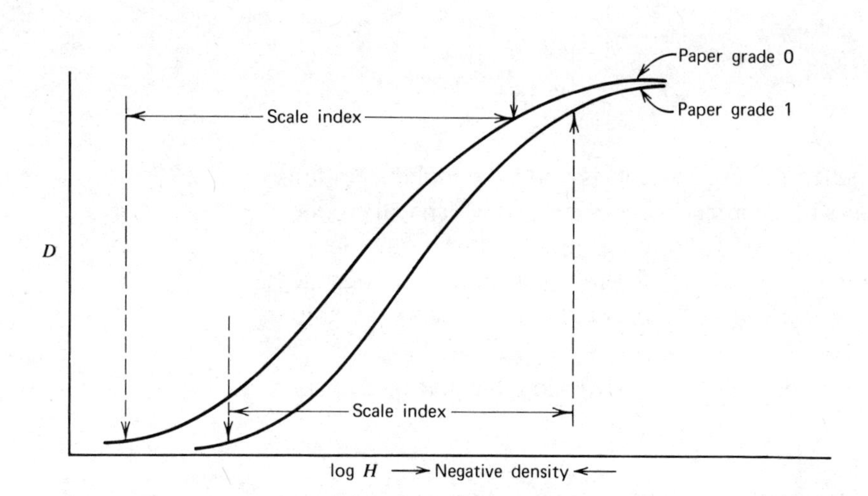

"Soft" papers can accommodate negatives having a ____________ density range. — large (etc.)

36. Papers called "hard" are graded 4 and 5; they have small-scale index values, as sketched in the graphs that follow.

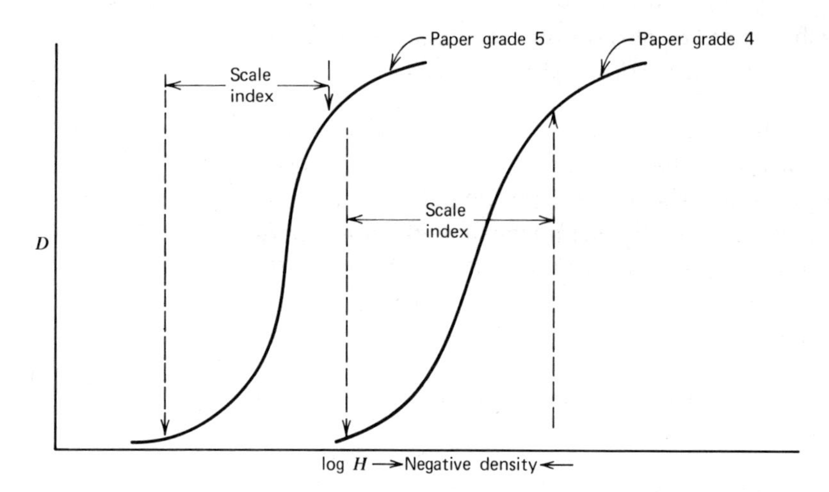

"Hard" papers produce good prints for negatives having a __________ density range. — small

37. For the two samples that you have the glossy paper has a __________ scale index (smaller, larger) than the mat paper. — smaller

38. By comparison with the mat paper, the glossy paper is __________ (softer, harder). — harder

39. Manufacturers of photographic papers publish approximate scale index values for the papers they sell. An example is the following:

Paper Grade	*Scale Index*
0	1.5
1	1.3
2	1.1
3	0.9
4	0.7
5	0.5

By using such a table (or, better, scale-index values you have found from tests described in this text) you can sort different negatives into sets which will produce reasonably good prints on a given paper. The rule, again, is that the negative density range must match the paper scale index. Using this rule, you know that a negative having extreme densities of 0.2 and 1.5 would require a paper of grade __________ . — 1

40. By the same rule, a paper having a density range from 0.4 to 1.1 would need a paper of grade __________ . — 4

The rule you have applied here is only an approximation. Different subjects and especially different tastes of the photographer or the client will no doubt demand exceptions to the rule.

SELF-TEST ON PHOTOGRAPHIC PAPERS—*D*–LOG *H* CURVES AND SCALE INDEX

Check your understanding of this chapter by answering the following questions. The correct answers follow.

Two different filters—#2 and # 5—were used in testing a variable-contrast paper. The exposure time was 16 seconds; a measurement of the light on the easel without the filters or step tablet gave a reading of 2.0 lux (meter-candles).

1. The exposure to the unobstructed light would have been __________ lux-seconds (meter-candle-seconds).
2. The log exposure under the same circumstances would have been __________ .
3. In the test, a standard negative step tablet was used, having a density range from 0.0 to 3.0 in steps of 0.15. The maximum log exposure received by the sample was __________ .
4. The minimum log exposure received by the sample was __________ .
5. In the following table you find the results of the experiment, omitting the log-exposure values that gave repeated values at the extremes of the step tablet.

LOG H	$\bar{1}.10$	$\bar{1}.25$	$\bar{1}.40$	$\bar{1}.55$	$\bar{1}.70$	$\bar{1}.85$	0.00	0.15	0.30	0.45	0.60	0.75	0.90	1.05
#2	0.00	0.02	0.08	0.21	0.42	0.66	0.93	1.24	1.54	1.76	1.88	1.91	1.94	1.97
#5			0.00	0.03	0.09	0.26	0.58	1.02	1.55	1.89	1.96	2.00		

Plot the two curves on the graph paper on page 124. The paper surface was __________ .

6. For the #2 filter, the minimum useful point has a log-exposure value of __________ .
7. For the same filter, the maximum useful point has a log-exposure value of __________ .
8. For the same filter, the value of the scale index is __________ .
9. For the #5 filter, the log-exposure value for the minimum useful point is __________ ; for the maximum useful point the log *H* is __________ ; the scale index is __________ .
10. To print a negative having a density range from 0.25 to 1.40, you would use the # __________ filter.
11. To print a negative having a density from 0.25 to 1.05, you would use the # __________ filter.

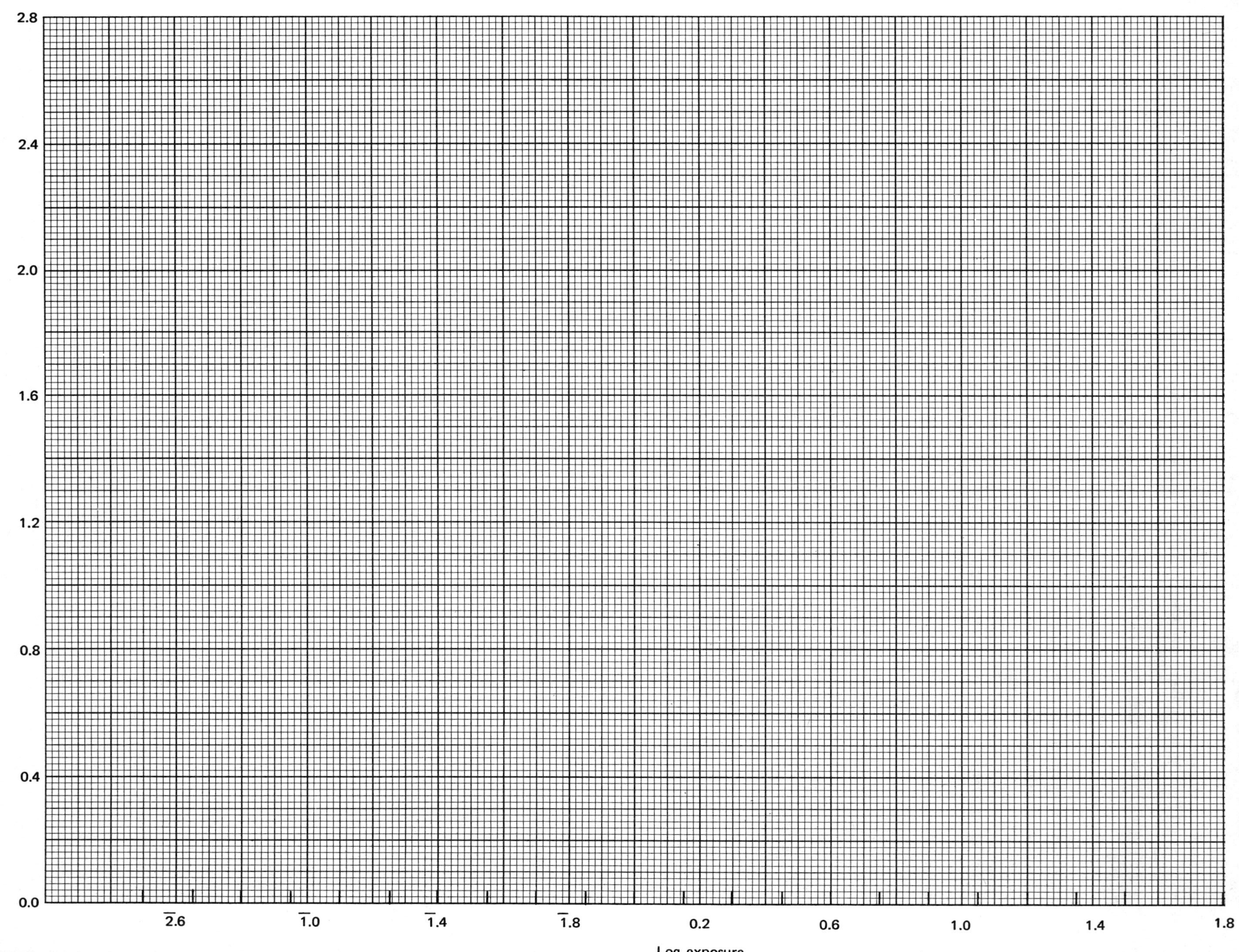
Density
2.8
2.4
2.0
1.6
1.2
0.8
0.4
0.0
$\bar{2}.6$
$\bar{1}.0$
$\bar{1}.4$
$\bar{1}.8$
0.2
0.6
1.0
1.4
1.8
Log exposure

ANSWERS TO SELF-TEST ON PHOTOGRAPHIC PAPERS—*D*–LOG *H* CURVES AND SCALE INDEX

In the parentheses after each right answer you find the items of this chapter that relate to each question and its answer.

1. 32 (item 1).

2. 1.50 (item 2).
3. 1.50 (items 3, 5).

4. $\bar{2}.50$ (items 4, 5).

5. A correctly plotted graph follows. The surface was glossy, based on the high value of *D*max (item 11).

6. $\bar{1}.32$ (items 22, 23, 29).

7. 0.48 (items 23–25, 30, 31).

8. 1.16 (items 25–26, 32–33).

9 $\bar{1}.60$, 0.38, 0.78 (items 22–26, 29–33).

10. #2 filter (items 26, 27, 34).

11. #5 filter (items 26, 27, 34).

NOTE. For practice in subtracting logarithms, as needed for test questions 8 and 9, see Appendix Section A5.

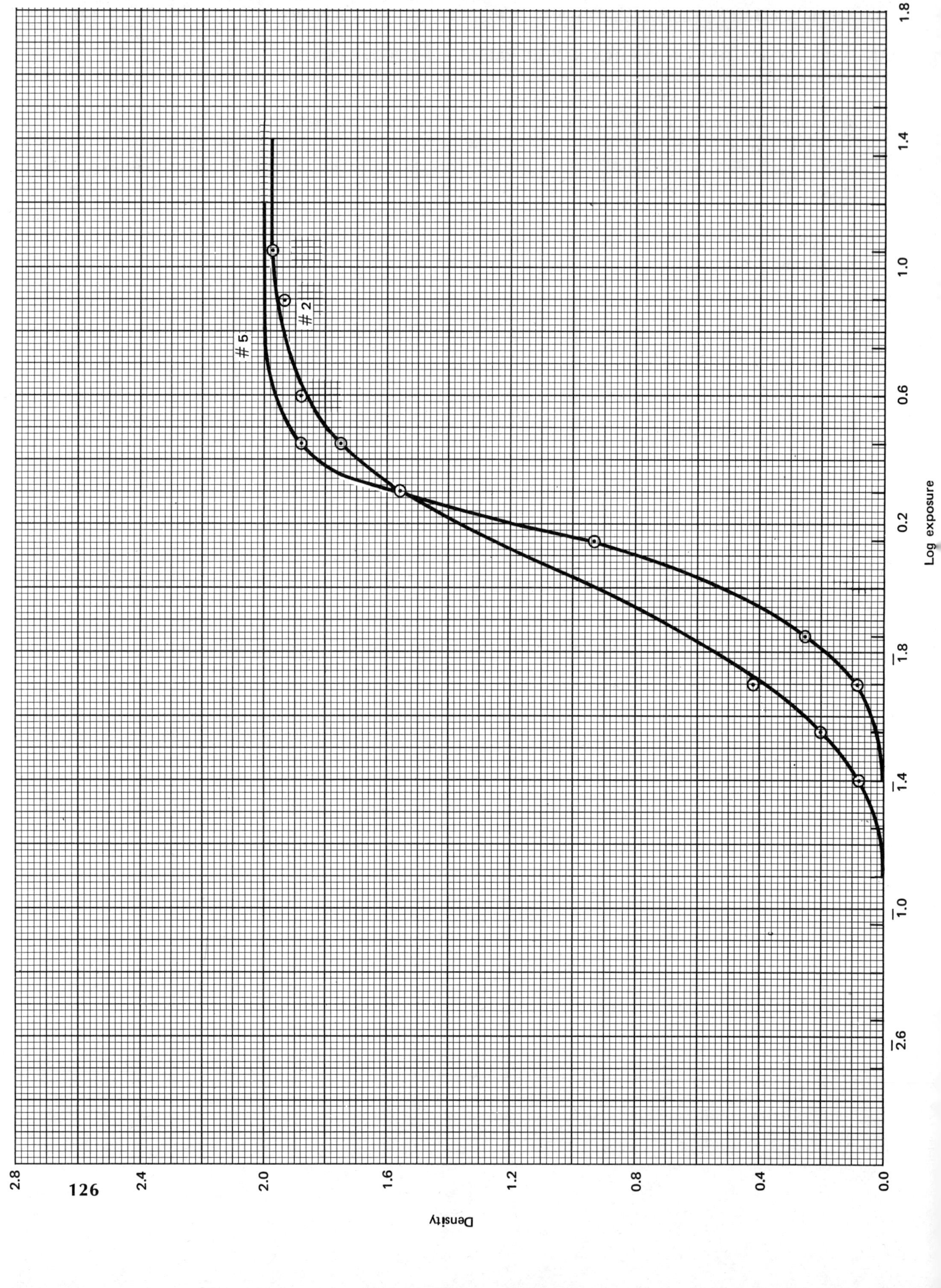

Density
2.8
2.4
2.0
1.6
1.2
0.8
0.4
0.0
#5
#2
$\bar{2}.6$
$\bar{1}.0$
$\bar{1}.4$
$\bar{1}.8$
0.2
0.6
1.0
1.4
1.8
Log exposure

CHAPTER 12

The Negative and the Print

In this chapter we apply the concepts of Chapter 11 to the process of making a print from a negative.

WHAT YOU WILL LEARN

If you work carefully with the material of this chapter, you will understand:

1. The necessary relationship between the negative and the printing paper for a full range of print tones;
2. The effects on the print of a change in exposure settings of the printer;
3. The effects on the print of a change in paper grade.

DIRECTIONS. Cover about 2 inches of the right-hand margin of each of the following pages in turn with an opaque sheet of paper. Read carefully the first numbered statement. *Write* the word or phrase that you believe correctly completes the statement. Move the cover sheet down to reveal the correct answer in the margin. Continue in this way until you have finished the section. Work at a rate that is comfortable for you, and no longer than you find pleasant. Begin when you wish.

1. The right relationship between the negative and the paper on which it is printed is shown in the sketch below.

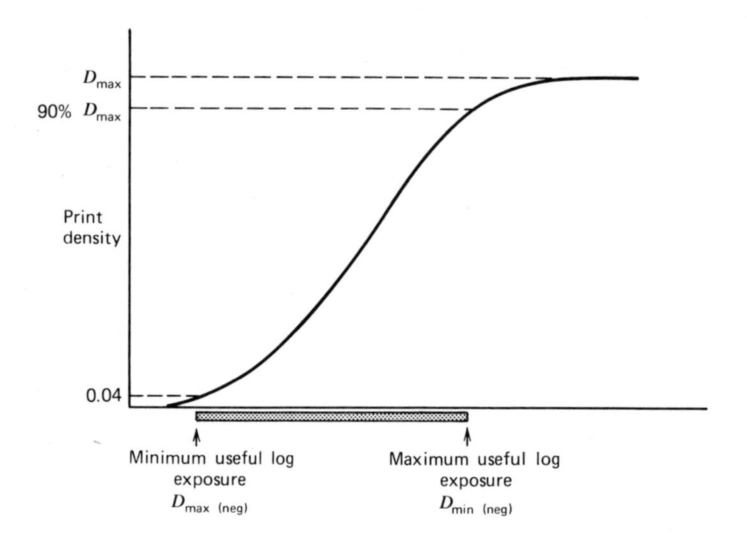

As you saw in the preceding section, the negative densities control the print exposures. Thus the left-hand end of the bar is labeled "minimum useful log exposure"; it corresponds to ____________ density present in the negative.
largest, smallest

largest

2. Therefore the left-hand end of the bar is also labeled $D_{max(neg)}$. The exposure level in the printer was adjusted so that this negative density produced an exposure on the print that gave the minimum useful print density, that is, a highlight density in the print of ____________ .

0.04

3. Similarly, the right-hand end of the bar representing the negative is labeled "maximum useful log exposure"; it corresponds to the ____________ density in the negative.

smallest

4. This density in the negative produced in the printer an exposure that gave the maximum useful density in the print, specifically, ____________ % of the maximum possible print density.

90

5. Thus the sketch shows the correct match between the negative density range and the ____________ ____________ of the paper.
(two words)

scale index

6. In the preceding sketch the *length* of the bar represents the density range of the negative, which in this case matches the scale index of the paper. Furthermore, the *placement* of the bar on the paper log *H* axis represents a correct print exposure level. *Over*exposure of the print is shown in the following sketch.

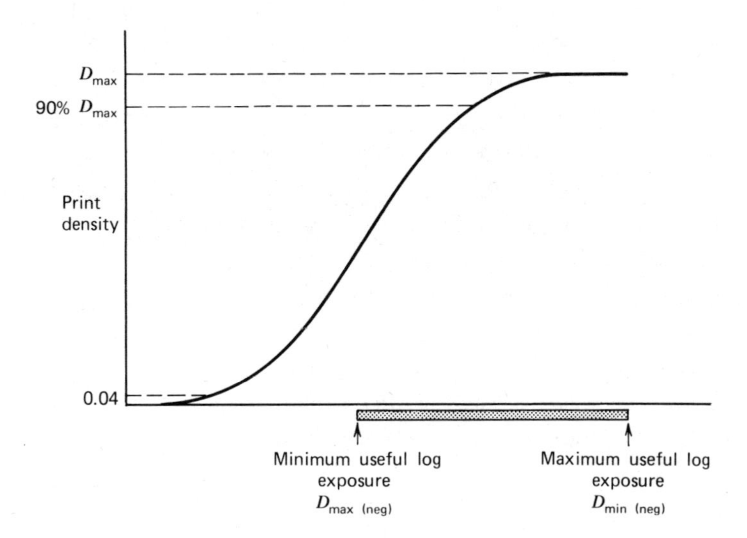

Now, the bar representing the negative has been moved to the *right,* in the direction of increased exposure for the print. Here the largest density of the negative produces an exposure so great that the smallest print density is well above the desirable value of 0.04, and the extreme highlight tone in the print would therefore be too __________ . — dark

7. What is worse, the smallest density in the negative produces an exposure for the print lying on the uppermost part of the paper curve where the slope is practically __________ . — zero
8. Zero slope means no __________ . — contrast or detail
9. Thus overexposure of the print causes a loss of contrast in the __________ . — shadows, or dark tones
10. We use a similar sketch to show *under*exposure of the print:

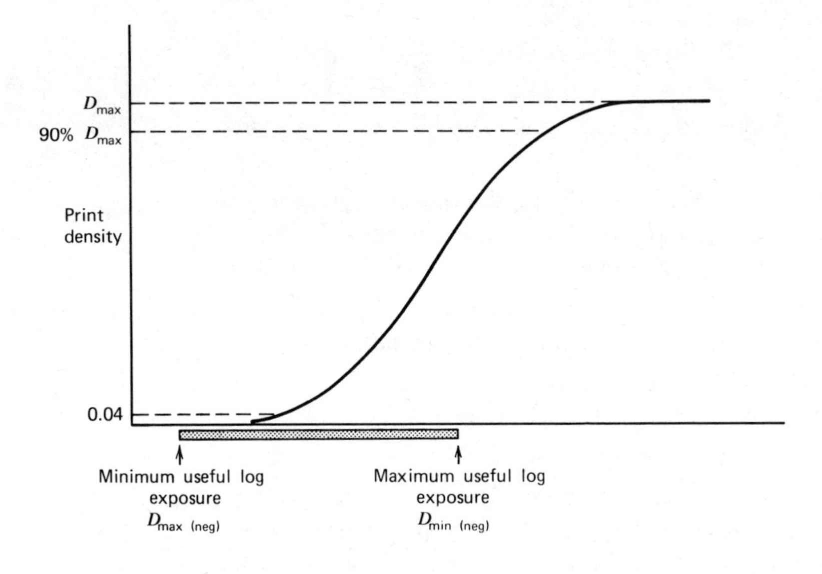

Here the bar representing the negative has been moved to the *left* on the print log-exposure axis. Now the D_{max} of the negative produces an exposure so small that the lightest tone in the print would have a density of ____________ . zero

11. Furthermore, the lightest tones of the print would have a slope of zero, and therefore no ____________ . contrast, or detail
12. Thus the underexposed print has "blocked-up" or "chalky" highlights. On the other hand, the darkest tone of the underexposed print would have a density much less than the preferred tone, and the dark (shadow) tones would be too ____________ . light
13. The preceding items involved an ideal situation in which the negative density range exactly matched the paper scale index. A mismatch, if only a slight one, is more likely. In the sketch that follows, we show a more realistic situation.

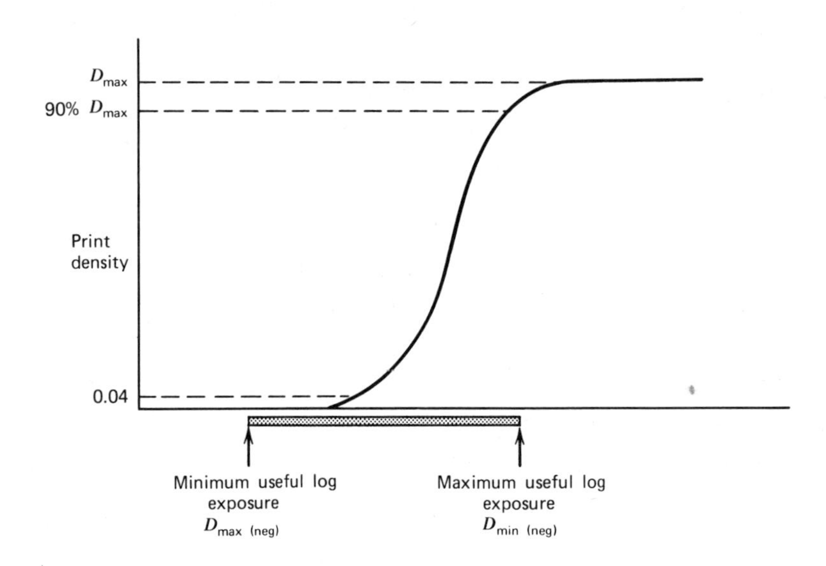

In this sketch, the paper scale index is ____________ than the negative density range. less, or smaller

14. This sketch also shows that the D_{min} of the negative produced a print exposure so that the dark tones of the print were all right. The D_{max} of the negative, however, was so great that the exposure it produced gave highlights at zero density and with no ____________ . contrast, or detail
15. The sketch represents a situation in which the paper is too "hard" for the negative; that is, the paper scale index is too ____________ . small
16. The exposure level was correct for the ____________ tones. dark, or shadow
17. The reproduction of the ____________ tones was poor. highlight
18. By changing the exposure level for the paper, a series of different prints could be made on the same paper. If the exposure level had been greatly increased, the situation would be as shown in the sketch that follows.

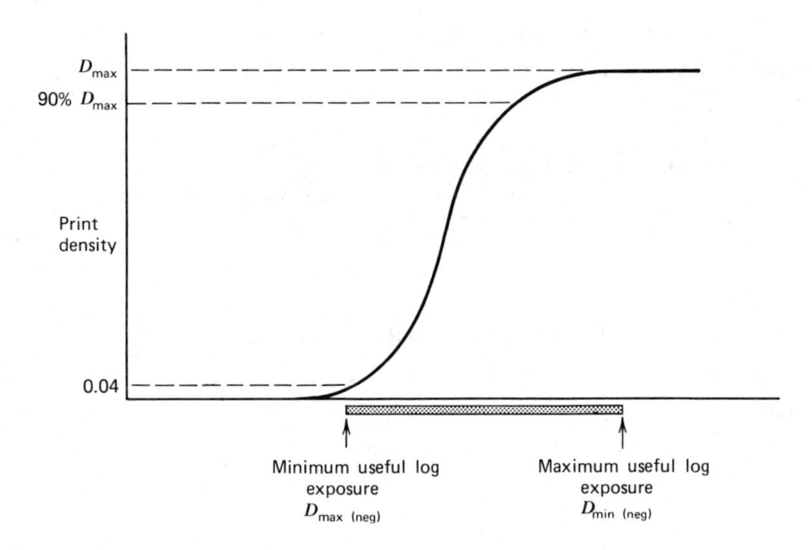

The bar representing the negative has been moved to the ________ .	right
19. Now, the exposure level is correct for the ________ .	highlights
20. Now, however, the D_{min} of the negative produces an exposure that is too great, and the dark tones of the print are too dark and have no ________ .	contrast, or detail
21. A print could be made with an exposure level intermediate between those shown in the two previous illustrations, in which the midtones would be correctly reproduced, but in which both the ________ and the ________ would be poor.	highlights, shadows (either order)

22. When the negative density range and the paper scale index are also mismatched, but the paper scale index is *greater* than the negative density range; the paper is too "soft" for the negative. Two of the many possible situations are sketched as follows.

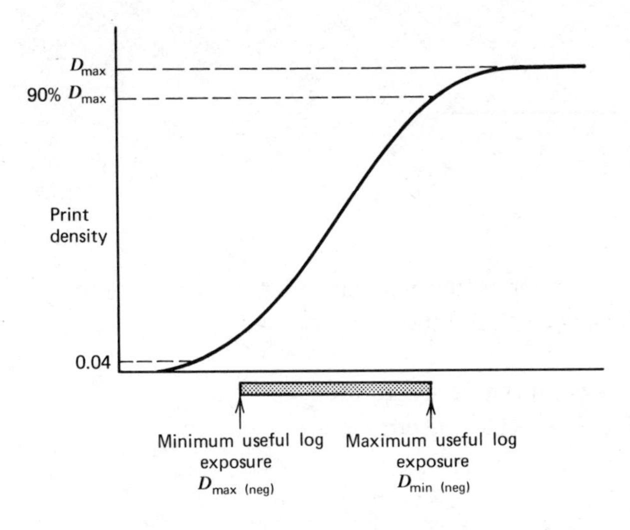

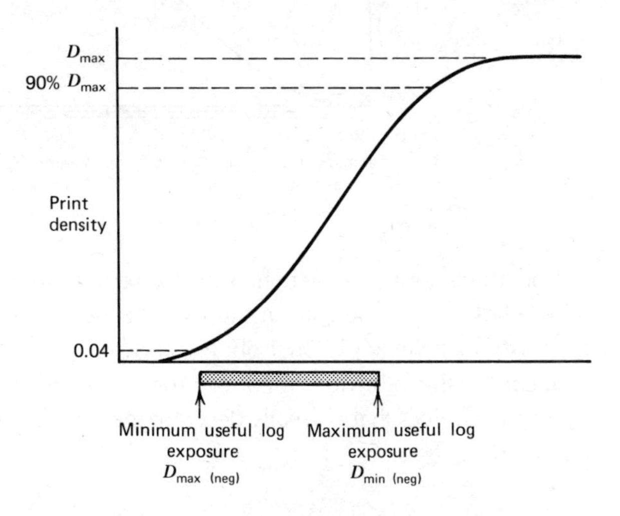

In these sketches the same negative was printed at two different exposure levels on a paper that was too soft. The exposure level was greater for the situation sketched at the _______ .
right, left

left

23. In the sketch at the left, correct reproduction is indicated for the _______
shadows,
_______ .
highlights

shadows

24. The highlights are poor; they are too __________ .

dark

25. With less of an exposure level, as shown in the sketch at the right, the __________ are reproduced correctly.

highlights

26. In the sketch at the right, the shadows are too __________ . Both of the prints made as sketched above would look too "flat"; a good print must have the highlights and shadows both reproduced correctly.

light

SELF-TEST ON THE NEGATIVE AND THE PRINT

Test your understanding of this chapter by answering the following questions. The correct answers follow. The sketch below duplicates the one in item 1 of this chapter.

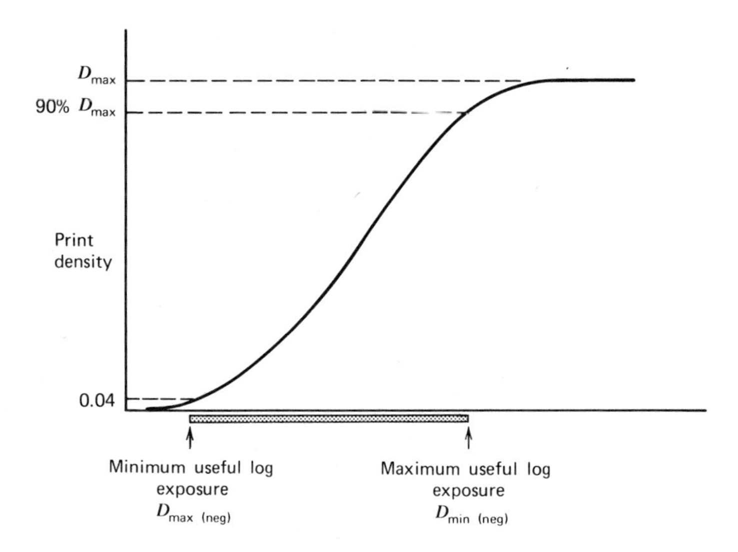

Both the shadows (at the right of the bar representing the negative) and the highlights (at the left) are reproduced in the print at the correct density level and the correct contrast, for general pictorial photography.

In each of the following cases, an error was made in making the print. For each case, select the choices that correctly describe the resulting print, and specify the error that was made.

1.

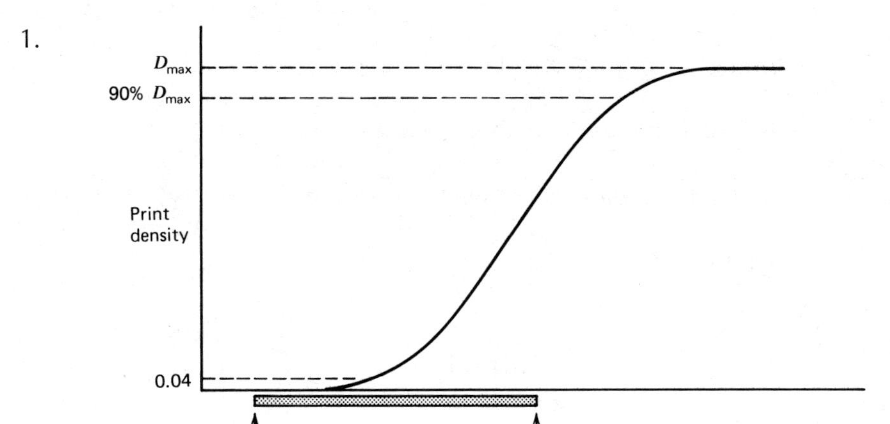

(a) The shadows are ____.
too light, too dark, about right in density

(b) The shadows are ____.
too flat, too contrasty, about right in contrast

(c) The highlights are ____.
too light, too dark, about right in density

(d) The highlights are ____.
too flat, too contrasty, about right in contrast.

(e) The error in making the print was:

____________ overexposure ____________ paper too hard
____________ underexposure ____________ paper too soft

2.

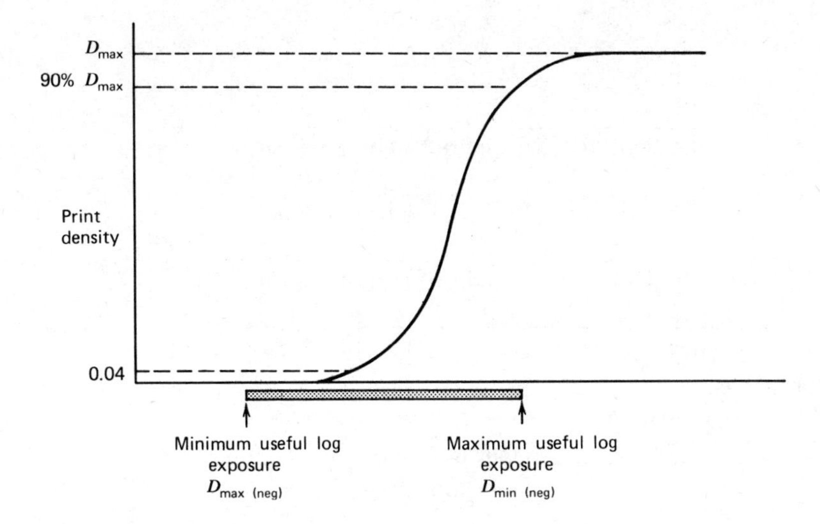

(a) The shadows are ______________________________.
too light, too dark, about right in density.

(b) The shadows are ______________________________.
too flat, too contrasty, about right in contrast.

(c) The highlights are ______________________________.
too light, too dark, about right in density.

(d) The highlights are ______________________________.
too flat, too contrasty, about right in contrast.

(e) The error in making the print was:

__________	overexposure	__________	paper too hard
__________	underexposure	__________	paper too soft

3.

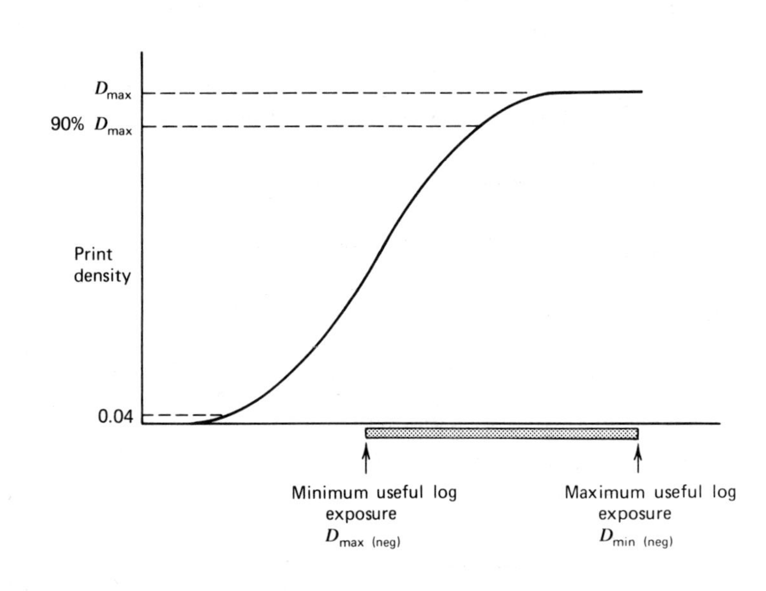

(a) The shadows are ______________________________.
too light, too dark, about right in density.

(b) The shadows are ______________________________.
too flat, too contrasty, about right in contrast.

(c) The highlights are ______________________________.
too light, too dark, about right in density.

(d) The highlights are ______________________________.
too flat, too contrasty, about right in contrast.

(e) The error in making the print was:

__________	overexposure	__________	paper too hard
__________	underexposure	__________	paper too soft

4.

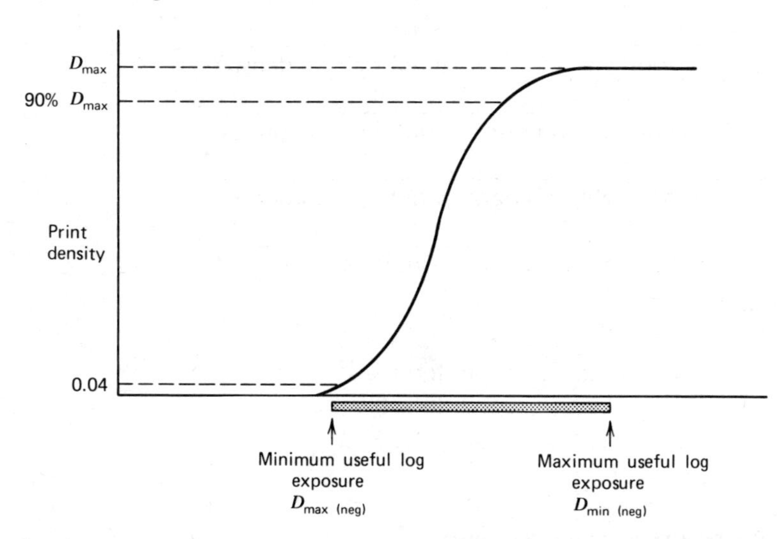

(a) The shadows are ______________________________.
too light, too dark, about right in density

(b) The shadows are ______________________________.
too flat, too contrasty, about right in contrast

(c) The highlights are ______________________________.
too light, too dark, about right in density

(d) The highlights are ______________________________.
too flat, too contrasty, about right in contrast

(e) The error in making the print was:

__________ overexposure __________ paper too hard

__________ underexposure __________ paper too soft

5.

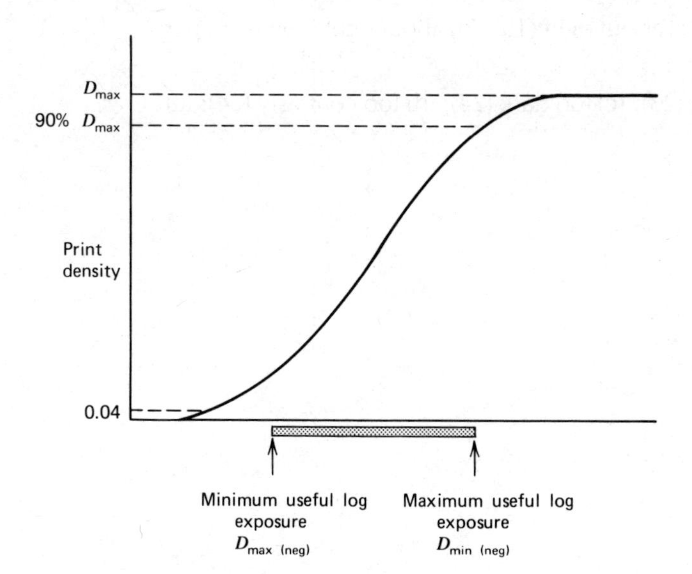

(a) The shadows are ______________________________.
too light, too dark, about right in density

(b) The shadows are ______________________________.
too flat, too contrasty, about right in contrast

(c) The highlights are ______________________________.
too light, too dark, about right in density

(d) The highlights are ______________________________.
too flat, too contrasty, about right in contrast

(e) The error in making the print was:

__________ overexposure __________ paper too hard
__________ underexposure __________ paper too soft

ANSWERS TO SELF-TEST ON THE NEGATIVE AND THE PRINT

In the parentheses after each answer you find the items in this chapter that relate to the question and its answer.

1. (a) Too light (12); (b) too contrasty—the shadows fall on a too-steep portion of the curve; (c) too light (10); (d) too flat (11, 12); (e) underexposure (10).
2. (a) About right (14); (b) about right (14); (c) too light (14); (d) too flat (14); (e) paper too hard (15).
3. (a) Too dark (the shadows fall too far up the curve); (b) too flat (7–9); (c) too dark (6); (d) too contrasty—the highlights fall on a too-steep part of the curve; (e) overexposure (6).
4. (a) Too dark (20); (b) too flat (20); (c) about right (19); (d) about right (19); (e) paper too hard (15).
5. (a) About right (23); (b) about right (23); (c) too dark (24); (d) too contrasty (24); (e) paper too soft.

CHAPTER 13

Paper Speeds and Exposure Determination

You saw from the section of this text dealing with the speeds of negative materials:

1. That a speed value is connected with an exposure method, or exposure system;
2. That a speed value is based on the exposure needed to produce a specified image.

Exactly the same concepts apply to speed values for photographic papers.

WHAT YOU WILL LEARN

If you work carefully with the following material, you will see that by knowing the speeds of photographic papers, you can:

1. Predict the needed change in exposure conditions when you change paper grade for a given negative;
2. From a single practical test, predict how to print any negative on any paper without further testing.
3. Once a print has been correctly exposed at one degree of enlargement, be able to make a differently sized print without further testing.

The methods of this chapter eliminate much of the waste of time and materials involved in trial-and-error procedures.

DIRECTIONS. Cover about 2 inches of the right-hand margin of each of the following pages in turn with an opaque sheet of paper. Read carefully the first numbered statement. *Write* the word or phrase that you believe correctly completes the statement. Move the cover sheet down to reveal the correct answer in the margin. Continue in this way until you have finished the section. Work at a rate that is comfortable for you, and no longer than you find pleasant. Begin when you wish.

From the extensive experimentation that led to the ASA speed method for negative pictorial films, we know that observers are very sensitive to the way *shadows* are reproduced in prints. They prefer prints in which the shadows are printed at very nearly the *maximum useful density*. You have seen that the maximum useful print density is 90% of the absolute maximum density the paper can give.

For this reason, it is sensible to define the paper speed on the basis of the exposure needed to give the maximum useful density in the print. Shadow speeds for papers are found as follows:

1. Find the log exposure (log H) needed to produce a density equal to 90% of the absolute Dmax. Suppose this log H for paper A to be 1.30. Find the antilog of 1.30. It is __________ . — 20
2. Divide the exposure just found into the arbitrary constant 10,000 and obtain __________ . The "shadow speed" of paper A is 500. — 500
3. Suppose paper B to require a log exposure of 1.00 to produce the maximum useful density. The exposure represented by this log is __________ . — 10
4. Divide the answer from (3) into the constant 10,000 and obtain the shadow speed of __________ . — 1000
5. The formula for the paper speed is: $S_p = 10{,}000/H_{\text{m.u.d.}}$ $H_{\text{m.u.d.}}$ stands for the *exposure* needed to produce the maximum useful density in the paper. If paper C takes a log exposure of 1.4 to produce 90% of paper Dmax, the formula gives this paper a speed value of __________ . — 400
6. In item 24 of Chapter 11 you found the maximum useful log exposure for the glossy paper to be 1.17. From this log exposure, you know that the shadow speed of the glossy paper is __________ . — 680
7. In item 31 of Chapter 11 you found that the mat paper needed a log exposure for the maximum point of 1.85. This log exposure implies a shadow speed of __________ . — 140

The following table summarizes the speeds for these papers:

Paper	*Shadow Speed*
A	500
B	1000
C	400
Glossy sample	680
Mat sample	140

Now we proceed to show how paper speeds may be used to predict exposure changes when the paper is changed for a given negative.

8. Assume that by trial we find that a negative gives a good print on paper *A*. A "good" print means correct reproduction of the shadows. The printing conditions were: *f*/5.6, 2x magnification; 20 seconds.

   We now intend to make a print of the same negative with paper *B*. We compare the speeds of the two papers by finding the ratio of the *first (A)* to the *second (B)*, that is, 500/1000, which reduces to ________ . — 0.5

9. We use this ratio (here, a fraction) as a multiplier with the original exposure time of 20 seconds and obtain ________ seconds. — 10

   This is a sensible answer: paper *B* has twice the speed of paper *A*, and logically takes half the time.

10. If we were to shift to paper *C*, which has a speed of 400, we again compare with paper *A*, and find the ratio of *A*/*C*, which reduces to ________ . — 1.25

11. We use the 1.25 as a multiplier with the original exposure time (20 seconds) and obtain ________ seconds. — 25

12. Follow the same procedure to predict the printing time of the same negative with the glossy sample. The time would be ________ . — 15 seconds

13. For the mat sample, the same negative would print well (for the shadows) at ________ seconds. — 71

NOTE. In these examples we have assumed a fixed *f*-number and magnification in the printer, and have been predicting the needed exposure time. This method works in the absence of reciprocity effects, in other words, when the exposure time does not need to be changed drastically. In published paper speed data, the exposure *time* is, in fact, fixed for the tests, and the illuminance instead is changed. For this reason, manufacturers' data will be more reliable if the printer *f*-number is changed, instead of the time.

SUMMARY. You have seen in Chapter 11 how to fit the paper to the negative if you know: (1) the paper scale index; (2) the negative D_{min} and D_{max}, from which you find the density range of the negative. Recall that (1) and (2) need to be equal. You have just seen how to figure out the needed change in exposure when you change papers, knowing the paper speed values.

We turn now to this problem: how to predict exposure conditions when the *negative* is changed.

14. We have been basing paper speeds on the right reproduction of the shadows. It is the corresponding shadow density in the negative D_{min} that gives us the key to the needed change in exposure conditions for different negatives.

    By trial we find that a negative with D_{min} of 0.40 gives good shadows on a given paper with a time of 20 seconds, other conditions being fixed. All negatives with that same D_{min} would print good shadows at the same time on the same paper. Another negative, with *denser* shadows, would need a ________ printing time. — longer, or greater

15. If the second negative has a D_{min} of 0.70, we find the needed time for that negative by first finding the difference between the D_{min} values for the first and second negatives. The difference is ________ . — 0.30

16. Since densities are logarithms, this difference in D_{min} values is also a log. We take the antilog of 0.30. The result is ________ . — 2

17. Now, the 2 just found is the *factor* by which the original exposure time (20 seconds) must be multiplied to get the time necessary for the second negative. The answer is __________ seconds. — 40

18. We intend to print a third negative on the same paper under the same conditions. Negative #3 has a D_{min} of only 0.10. It will take an exposure time, as compared with the first (item 14) that will be __________ . (less, the same, more) — less

19. Again, the clue is the difference between the D_{min} values of #3 and #1 in item 14. It is __________ . — 0.30

20. Here, since negative #3 is *thinner* in the shadows, we use the antilog of 0.30 (which is again 2) as a *divisor* with the original exposure time of 20 seconds and obtain __________ seconds. — 10

    The principles are these: we do a practical test with a reference (or standard) negative. We need to know the D_{min} values for the reference and any other negative. The difference in the D_{min} values, converted to the antilog, gives us a factor. We use that factor as a *multiplier* with the exposure time if the sample negative is *denser* than the reference, and as a *divisor* if the sample is *thinner*.

21. Comparing yet a fourth negative with the standard, if negative #4 has a D_{min} of 0.55, this is __________ more than the D_{min} of the first (reference) negative in item 14. — 0.15

22. The antilog of 0.15 is __________ . — 1.4

23. The reference needed a time of 20 seconds. Under the same conditions, negative #4 will take __________ seconds. Negative #4 is denser in the shadows than the reference, so the factor 1.4 is used as a multiplier. — 28

24. Negative # 5 has a D_{min} of 0.20. It differs from the reference by a value of __________ in the shadows. — 0.20

25. The factor is __________ . — 1.6

26. The exposure time under the same conditions would be __________ seconds. Negative # 5 is thinner than the reference in the shadows, so the number 1.6 is used as a divisor. — 12–13

In this section we have shown how to do exposure calculations separately for: (1) paper speeds and (2) negative D_{min} values. By doing two successive calculations you can take both of these factors into account. Thus having made tests on papers to find their speeds, and having made just one practical test with one reference negative, you can thereafter (with a stable process) figure out the exposure conditions needed for *any* negative printed on *any* paper. In the following we illustrate the method.

We have tested a variable contrast paper, and obtained the following:

Filter	*Scale Index*	*Shadow Speed*
1	1.5	250
2	1.3	200
3	1.1	100
4	0.9	80

We have found by trial the necessary exposure conditions for a reference negative:

Negative (reference)	*Exposure conditions*
$D_{min} = 0.25$; $D_{max} = 1.55$	10 seconds, *f*/4.5, 2x magnification

27. Since the reference negative had a difference of 1.3 from its D_{min} to its D_{max}, from the rule that the paper scale index needs to match the negative density range, the reference negative would have been printed with filter # ________ . — 2
28. We now want to print negative *A*, which has a D_{min} of 0.25 and a D_{max} of 1.35. Since its D_{min} matches that of the reference, we need not concern ourselves with that. The density range of the negative, however, is ________ . — 1.10
29. Matching the negative density range with the paper scale index, we see that negative *A* will require a # ________ filter. — 3
30. We now compare the speeds of the paper when exposed with the two filters (#2 vs. #3), and using the methods of items 8–13 above we arrive at a needed exposure time of ________ seconds. — 20
31. Consider now negative *B*. It has a D_{min} of 0.55 and a D_{max} of 1.65, for a range of ________ . — 1.10
32. Negative *B* will require a # ________ filter. — 3
33. Considering first only the filter change (ignoring for the moment the D_{min} difference) a time of ________ seconds would be needed. — 20
34. In addition, however, we need to compensate for the increase in the D_{min} of negative *B* as compared with the reference. The difference between the D_{min} values is ________ . — 0.30
35. The antilog of 0.30, 2, is a factor that must be applied to the exposure time found in the answer to item 33.
36. Since negative *B* is *denser* than the reference negative, 2 must be used as a *multiplier* with the 20 seconds (item 36) to give the final answer of ________ seconds. — 40

When a sample negative differs from the reference both in density range and in D_{min}, we compute the needed exposure time in two stages—first considering the paper-speed ratio, and then the D_{min} difference from the reference. Thus we find two factors, and use them in sequence.

37. Negative C has a D_{min} of 0.20 and a D_{max} of 1.10. From the range of this negative, it will need a # ________ filter. — 4
38. We find the ratio of the paper speeds to be ________ . — 2.5
39. We use this as a multiplier with the time used for the reference negative, and have the first stage of the answer: ________ . — 25
40. An exposure time of 25 seconds would be correct if negative *C* had the same D_{min} as the reference. But, in fact, negative *C* is slightly thinner in the shadows; its D_{min} is ________ less than that of the reference. — 0.05
41. The antilog of 0.05 is ________ . — 1.12
42. Since the sample negative is thinner than the reference in the shadows, we use the number 1.12 as a divisor with the partial answer (25 seconds) and obtain ________ seconds. — 22
43. In items 37–43 we have compensated for both the change in filter and the change in the negative D_{min}, by dividing the problem into two parts. Finally, consider

negative *D*, having a D_{min} of 0.15 and a D_{max} of 1.65. A # __________ filter will be used. | 1

44. The change in paper speed alone would call for a change in exposure time of 10 seconds used for the reference to __________ seconds for negative *D*. | 8
45. But also negative *D* is thinner in the shadows than is the reference. The difference between D_{min} values is __________ . | 0.10
46. This change further reduces the 8 seconds to a little more than __________ seconds. | 6

To complete this discussion of exposure determination for a projection printing situation, we consider now the needed change when the degree of enlargement of the negative is altered. If the correct exposure settings (*f*-number and time) are known for a 4 × 5 print, we can know without further tests how to make an 8 × 10 print from the same negative. Unfortunately, a common-sense approach usually leads to a badly exposed print. Specifically, the "inverse-square" law (Appendix D, items 9–18) will not work in this situation. The law applies only to a *point* source, which is certainly not true for a projection printer. We develop the correct method in the following items.

47. The clue to the needed change in exposure time is a comparison between the degree of enlargement in the first case and that in the second. In optics, "degree of enlargement" is called the *magnification*. A better term is "scale of reproduction." *Scale of reproduction* is the ratio of a linear dimension in the *image* compared with the corresponding dimension in the *negative*. For example, if one side of a negative measures 4 inches in the image and 1 inch in the negative, the scale of reproduction (or magnification) is __________ . | 4×
48. If the image of a person is 2 cm long in the negative and 6 cm long in the print, the magnification is __________ . | 3×
49. If the height of a building in the negative is 10 cm and in the print is 2.5 cm, the scale of reproduction is __________ . | ¼×, or 0.25×
50. The needed exposure time in making a projection print is proportional to the quantity $(R + 1)^2$, where *R* is the scale of reproduction.
    If a correctly exposed print is made at a scale of reproduction of 2×, we must add 1, to get 3, and then square the result to get __________ . | 9
51. If a second print is to be made at a magnification of 4×, we again add 1 and square the result to get __________ . | 25
52. The ratio of the second value (25) to the first (9) gives a number that must be used as a multiplier with the original exposure time to find the time needed for the second print. If the first print was correctly exposed at 10 seconds, the second will be correctly exposed for a time of 25/9 × 10, or __________ seconds. | 28
53. Assume that we have made a correctly exposed 5 × 7 print from a 2¼ × 3¼ negative at 5 seconds at *f*/5.6. The scale of reproduction is __________ . | 2.2×
54. We want to predict the needed exposure time at the same lens aperture to make a 16 × 23 print. The magnification in the second case is __________ . | 7.1×
55. We compute the term $(R + 1)^2$ for the first print (item 53). It is __________ . | 10.25
56. We compute a similar term for the second (desired) print. It is __________ . | 65.6
57. The ratio of the result in item 56 to that in item 55 is __________ to 1. | 6.4

58. We use this result as a multiplier of the original exposure time (5 seconds) to obtain the answer of __________ seconds.

32

SELF-TEST ON PAPER SPEEDS AND EXPOSURE DETERMINATION

Check your understanding of this chapter by answering the following questions. The correct answers follow.

1. If the log exposure for the maximum useful point on a paper *D*–log *H* curve is 1.60, the shadow speed from the formula $S_p = 10{,}000/H$ is __________ .

2. Two papers, *X* and *Y*, have shadow speeds of 200 and 500. If a negative is correctly exposed on paper *X* at a 10-second exposure time, for the same negative paper *Y* would require an exposure time of __________ seconds.
3. Papers *Q* and *R* have shadow speeds of 300 and 100. If a 20-second exposure time is needed for a given negative printed for the shadows on paper *Q*, paper *R* would need an exposure time of __________ seconds.
4. A standard negative, with a density range from 0.2 to 1.7, prints well at *f*/8 and 15 seconds. Another negative to be printed on the same paper, has a density range from 0.5 to 2.0. At *f*/8 for the second negative the exposure time would need to be __________ .
5. Refer to question 4. A still different negative has a density range from 0.1 to 1.7. Printed on the same paper, this negative would take a time of __________ seconds.
6. A 4× enlargement from a 35mm negative requires 4 seconds at *f*/5.6. For the same *f*-number, a 12× enlargement of the same negative on the same paper would need an exposure time of __________ seconds.

ANSWERS TO SELF-TEST ON PAPER SPEEDS AND EXPOSURE DETERMINATION

In the parentheses after each answer you find the items of this chapter that relate to each question and its answer.

1. 250 (items 1–6).
2. 4 (items 8–13).
3. 60 (items 8–13).
4. 30 seconds (items 14–17).
5. 12 (items 14–17).
6. 27 (items 47–58).

CHAPTER 14
Tone Reproduction

Many, perhaps most, photographs are made with the intent of showing a viewer how the subject of the photograph *looked*. The extent to which this intent is successful can be shown by the relationship between measurements of the subject and of the final image. The following section of this text deals with this relationship.

There are two somewhat different cases. In one case, what is wanted is *facsimile* reproduction; the image should, for complete success, duplicate the original. An example is in copying a line drawing or other graphic work. Another is reproducing a photograph in a magazine.

A somewhat different type of photography is involved when the photographer wishes to show the viewer how the White House or the Grand Canyon appeared at a particular time, or he wants to make a portrait of a person. It is not immediately obvious how the best photographic image will relate to the original subject, and, in fact, there is good evidence that facsimile reproduction is neither possible nor desirable, as will be seen later. But *some* relationship between photograph and subject is plausible.

The study of "tone reproduction" provides a method of describing such a relationship and thereby answering the question, "how ought the print, if it is to be of high quality, compare with the subject?"

(Note that there are other, equally valid, kinds of photography where the intent of the photographer is different. He may want to display—not an imitation of the original, but an abstraction or variation or interpretation of the original. For such photography, taste and esthetics far outweigh any considerations of fidelity or accuracy. Taste is a personal matter and not subject to the kind of treatment to be developed here.)

In summary, the method of tone reproduction is valuable as a means of studying the way in which a photographic process handles the subject. The result of the study is a graphical representation of the relation between the image and the original scene. In this section we consider black-and-white photography where a print is the final image.

A tone reproduction study requires data about the subject—luminance values, as in the items in Chapter 2, items 58–92 and Appendix C2. It also requires data about the print—reflection densities (see Chapter 2, items 34–50).

What follows assumes that the reader is familiar with the material referred to above and also with the basic concepts of logarithms (Appendix Sections A1–A4).

WHAT YOU WILL LEARN

If you work carefully with the following material, you will be able to:

1. Obtain useful numerical data about the subject tones;
2. Know from such data when normal methods of photography can be expected to produce good results;
3. Relate print-density measurements to print tones;
4. Make and interpret plots showing the relationship between the subject and the print.

DIRECTIONS. Cover about 2 inches of the right-hand margin of each of the following pages in turn with an opaque sheet of paper. Read carefully the first numbered statement. *Write* the word or phrase that you believe correctly completes the statement. Move the cover sheet down to reveal the correct answer in the margin. Continue in this way until you have finished the section. Work at a rate that is comfortable for you, and no longer than you find pleasant. Begin when you wish.

Subject Data

1. When we look at any ordinary subject, we see many different shapes, colors, and *tones*. A tone is one of a set of different lightnesses. A tone is called a "zone" in the zone system of Adams and White. A shadow usually has a dark __________ . — tone
2. In a sunny summer landscape including clouds and sky, the lightest tone would be associated with the __________ . — clouds
3. If the subject of a portrait were a brunette Caucasian dressed in gray clothing, the darkest tone would be found in the person's __________ . — hair
4. The lightest tone would be the person's __________ . — skin, or face
5. The gray clothing would have medium __________ . — tones
6. In black-and-white photography, think of a subject as a set of tones—dark tones for dark areas in shadow, light tones ("highlights"), and, in between, "midtones"

of medium lightness. A successful pictorial or documentary photograph must adequately "reproduce," or simulate, all of the important ________ .

tones

7. We can measure the tones of a subject with a meter having a narrow angle of view. The technical name for such a meter is ________ meter.

luminance

8. If a luminance meter is held so that it "sees" light from a single area of a subject, we obtain a reading that is related to the subject ________ .

tone

9. Our eyes respond nearly proportionally to the logarithm of the subject luminance. For this reason, most modern luminance meters are provided with scales related to logs. A portion of such a scale is shown in this diagram. One scale interval is

1 2 3 4 5 6 7 8 9 10 11

equivalent to 0.30 in logs. Thus if one tone of a subject gives a scale reading of 5 and another of 6, we know that the logs of the subject luminances differ by ________ .

0.30

10. A difference of 0.30 in logs corresponds to a multiplication of the corresponding numbers by ________ .

2

11. Thus if the scale readings are 5 and 6, we know that the luminance value for the area measuring 6 is ________ times that for the area measuring 5.

2

12. Every additional scale value represents an addition in logs of ________ .

0.30

13. Thus if two tones of a subject read 7 and 9, the total log interval is ________ .

0.60

14. The antilog of 0.60 is the factor relating the two subject luminances. The antilog of 0.60 is ________ .

4

15. If two subject tones measure 10 and 12, the one measuring 12 would have ________ times the luminance of the one measuring 10.

4

16. Suppose that in a portrait you want one side of the face to have twice the luminance of the other—a 2:1 ratio. If one side meters 10, the other (lighter) side must meter ________ .

11

17. If you want a 4:1 luminance ratio, and the lighter side of the face measures 8, the darker side must measure ________ .

6

18. Two subject tones measure 8 and 14. The number of scale intervals is ________ .

6

19. Since each of the scale intervals is 0.30 in logs, the total log interval represented by the values in item 18 is ________ .

1.80

20. The antilog of 1.80 is a little over ________ .

60

21. Thus if the meter readings are 8 and 14, the lighter tone is a little over ________ times the darker in luminance.

60

22. With care, you can estimate scale readings between whole-scale values. Each whole interval is divided into thirds by tick marks. Since the whole interval is 0.30 in logs, each tiny interval is ________ in logs.

0.10

23. If the meter readings for two subject tones were 5 1/3 and 7, the interval in scale readings would be ________ .

1 2/3

24. The whole interval is worth 0.30 in logs; the 2/3 is worth another 0.20, for a total log interval of ________ .

0.50

25. If two subject tones meter 3 2/3 and 8, the interval in scale readings is __________ . — 4 1/3

26. The log interval for item 25 is __________ . — 1.30

27. The antilog of 1.30 means that the lighter tone has __________ times the luminance of the darker tone. — 20

28. If the lightest and darkest tones of a subject are identified and measured, the corresponding log difference is a useful measure of the *total contrast* of the scene. We know from many measurements of outdoor subjects that the average value for the total contrast is about 2.2 in logs. This is slightly more than __________ meter divisions. — 7

29. Any outdoor scene measuring about 7 divisions from darkest shadow to lightest highlight can be considered typical. Rarely, an outdoor subject will give a range of as much as 3.0 in logs, or __________ scale divisions. — 10

30. A scene having such a range is very contrasty. An example of such a subject would be a winter snow scene including dark trees in bright sunlight. Conversely, a very "flat" outdoor scene may give a range of only 5 scale divisions, a total contrast in logs of only __________ . — 1.5

31. We can classify scenes into types—typical, contrasty, or flat—merely by finding the range of the subject tones in scale divisions, multiplying the value by 0.30, and comparing the result with the values in items 28–30. If the smallest reading is 5 and the largest is 13, the scene is close to __________ . — typical

32. Extreme readings of 4 and 10 would cause us to conclude that the scene is __________ . — flat

33. Readings of 5 and 15 1/3 would indicate that the scene is __________ . — contrasty

34. Simply knowing the total contrast of a scene often lets the photographer know whether or not his usual photographic process will succeed. Usual methods have been designed to give good results with nearly typical subjects. Usual methods fail (or must be greatly changed) to produce good images of scenes that are unusually flat or contrasty. For example, the ground as seen from an airplane may (because of the effects of the atmosphere) measure only two scale divisions between the darkest and lightest tones. Such a subject is very __________ . — flat

35. With ordinary methods, you cannot take a good photograph of such a flat scene. Conversely, a sunset scene that includes shadowed foreground and brightly lit clouds is very __________ . — contrasty

    To make a black-and-white print that properly represents the scene (as distinct from being a pleasing picture) is impossible.

36. You have seen that just two readings of a scene allow the photographer to make a good judgment about the total contrast of a scene. For tone-reproduction studies, at least a few more readings of intermediate tones are useful. The sketch that follows suggests an outdoor scene for which several meter readings have been made and marked on the sketch.

The range of readings from the value for the tree trunk to the cloud above indicates that the contrast of this scene is __________ . typical

37. The range of the scene in logs is __________ . 2.1
38. Think of each measured tone of the subject as representing one *input* value, in the sense that each scale reading is a relative measure of the light sent from the specific tone toward the camera. Thus we can make an ordered scale of input values on a relative basis, recalling that the log interval for one scale division is __________ . 0.30

The input scale is shown in the sketch that follows.

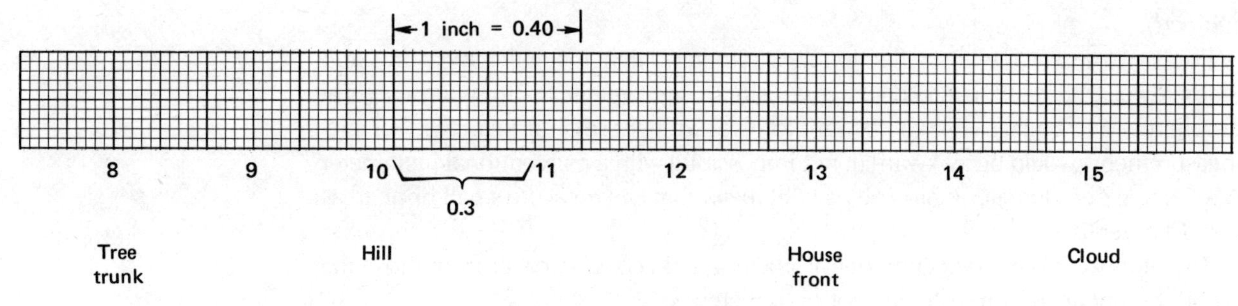

39. Note that the scale is labeled *relative* because only the intervals are of interest. It is a *log* scale, in fact, because each such interval is "worth" __________ in logs. 0.30
40. Using graph paper subdivided into inches, we have chosen to assign a value of 0.40 to each inch division. Using such a value for each inch, the 0.30 log interval between whole-number meter readings becomes __________ inch. 3/4
41. For the scene we have been considering, the meter readings were what they were because of the reflection characteristics of the objects in the scene and also the lighting. On an overcast day, for example, the meter readings would be different from those above. On such a day there is much less light than on the sunny day that we have been using as example and thus all the readings would be __________ than before. less

42. Also, because on an overcast day the lighting on the scene would be much more nearly uniform than on a sunny day, the range of the subject tones would also become ____________ than before.

less

43. Assume these readings on a dull day: sky, 11; hill, 8; tree trunk, 6; house front, 10; house side, 9. Place these values on the following scale, using the one there as reference.

9

House side

You find below a scale correctly arranged.

6	7	8	9	10	11
Tree trunk		Hill	House side	House front	Sky

Print Measurements and the Tone Reproduction Plot

If we have recorded subject data like those in the section above, we can now, by measuring the final print, answer the question: How does the print relate to the subject?

Prints are measured with a reflection densitometer, a light meter that measures print tones in logarithms. You could imagine using a light meter for this purpose. The problem would be that you would want to measure very small print areas, but such measurements would be awkward (if not impossible) with a conventional light meter. A reflection densitometer is basically a light meter that can measure small print areas. (See Chapter 10.)

The numbers from a reflection densitometer are *reversed in order* from those that would be obtained from a small-spot light meter.

44. We usually adjust the densitometer so that it reads zero when placed over the lightest tone the paper can produce. Thus an unexposed fixed-out paper area has a density of ____________ .

0

45. As the tone of the print becomes darker, the measured density becomes ____________ .

greater, or more

46. Thus a middle tone of a print may have a density of 1.0. The larger the density, the ____________ the tone of the print.

darker

47. For a glossy print, the darkest possible tone may have a density over 2, perhaps as great as 2.2. Recall, however, from Chapter 9, that the greatest *useful* density in a print is 90% of *D*max. Thus the greatest useful density in a glossy print is approximately ____________ .

2.0

48. We intend to plot a graph relating print measurements to subject measurements. You have seen in the preceding section how to place the subject data on a horizontal scale. We will place the print-density values on a vertical scale.

If we order the print-density values in the usual way, with the numbers increasing upward, the result will be that light tones of the print will plot near the ____________ of the scale and dark tones, near the ____________ .

bottom, top

49. For the print-density scale, we will use the same scale interval as on the horizontal scale, with 1 inch = ____________ .

0.40

You find the layout of the graph in the following diagram.

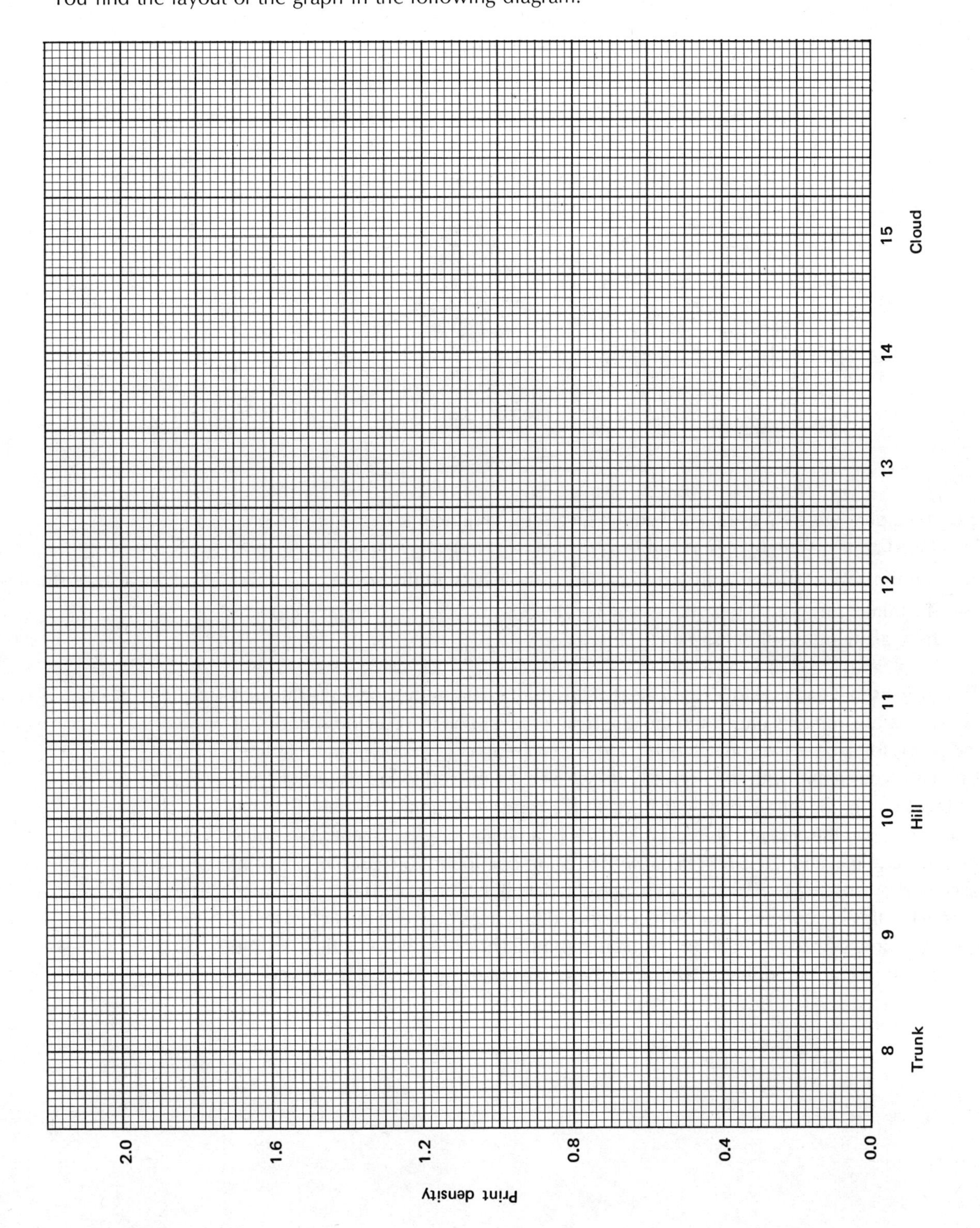

50. In this layout, note that both scales are logarithmic, with the horizontal axis showing data obtained from measurements of the ____________ . — subject
51. The subject data are the same as those in items 36–39 above. The value assigned to one inch of the graph paper is the same on both scales, that is, ____________ . — 0.40

We repeat below the sketch of the scene shown in item 36, but now the sketch represents a *print* of the subject. In parentheses are the subject meter readings. Below each meter reading you see the corresponding print density.

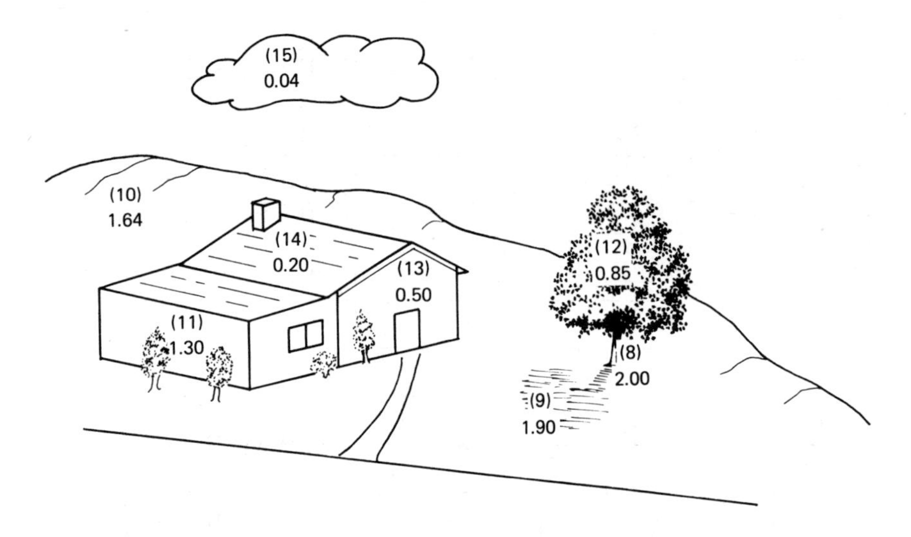

52. The cloud, which in the scene had the greatest luminance, is in the print reproduced at nearly the lightest possible tone, that is, at a density of ____________ . — 0.04
53. The trunk of the tree had the ____________ luminance in the scene. — lowest, or smallest
54. The trunk in the print is reproduced at a density of ____________ , which is close to the maximum useful density for a glossy print. — 2.00

The answers to items 53 and 55 indicate that this print contains the complete range of tones possible from this paper. On the graph paper layout on the previous page, plot the points for each of the subject tones, and with a French curve draw a smooth line. Compare your graph with the correctly drawn one on the following page.

The tone-reproduction plot you have made represents approximately the one obtained when the *best possible* print is made of a typical outdoor scene. We know that this curve is about the best because it was the one found to represent the print preferred by observers when they were presented with a variety of prints of the same scene. For this kind of photography, even small differences between the actual result and the curve shown above indicate inferior prints. Thus the interpretation of tone reproduction graphs requires attention to the details of the *shape* of the curve.

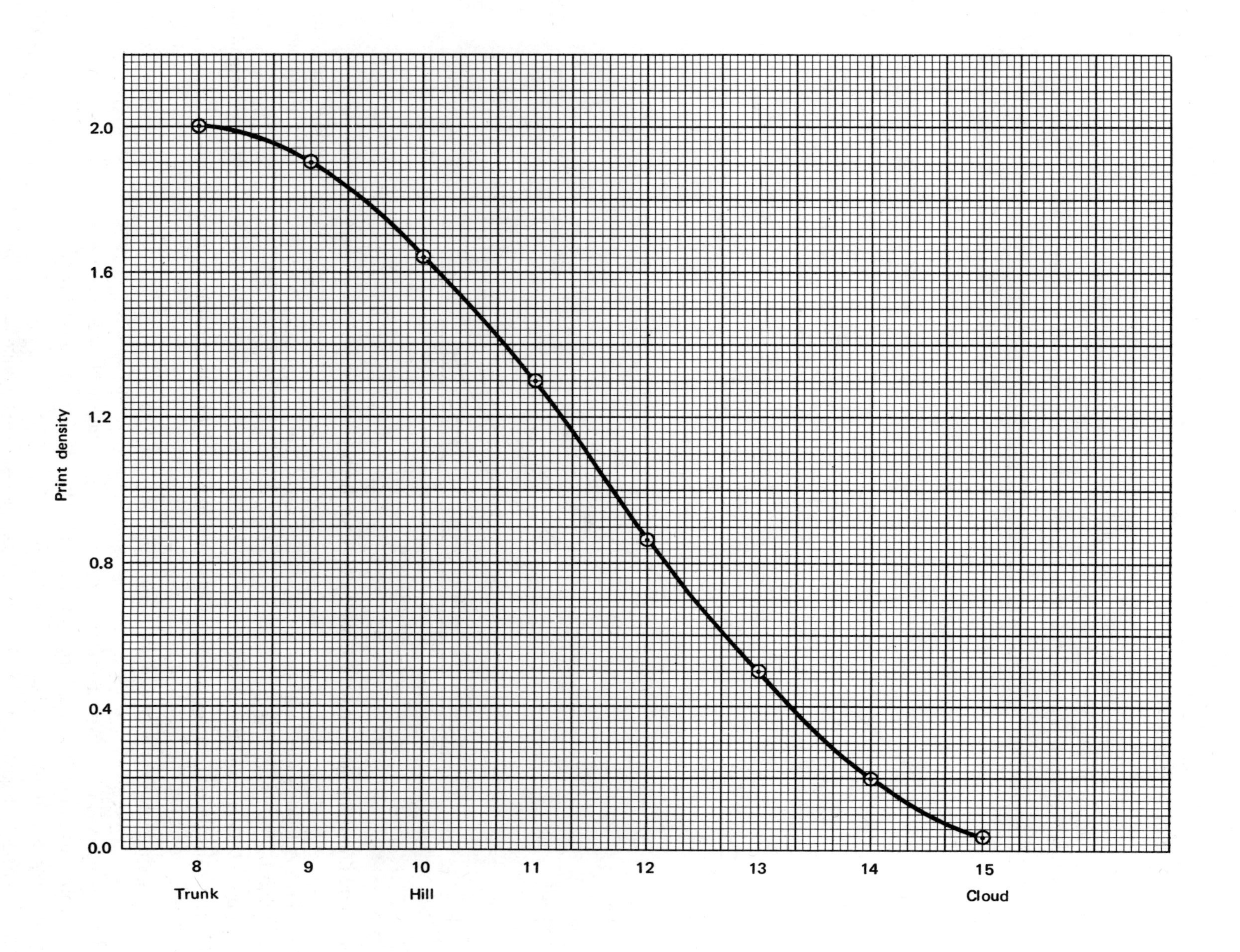

Print density
2.0
1.6
1.2
0.8
0.4
0.0
8
9
10
11
12
13
14
15
Trunk
Hill
Cloud

Interpretation of Tone-reproduction Plots

Significant differences between the aim curve and typical curves indicating less-than-best results are shown in the following items. We are invariably assuming that a glossy print has been made of a typical outdoor subject. Refer to the graphs that follow.

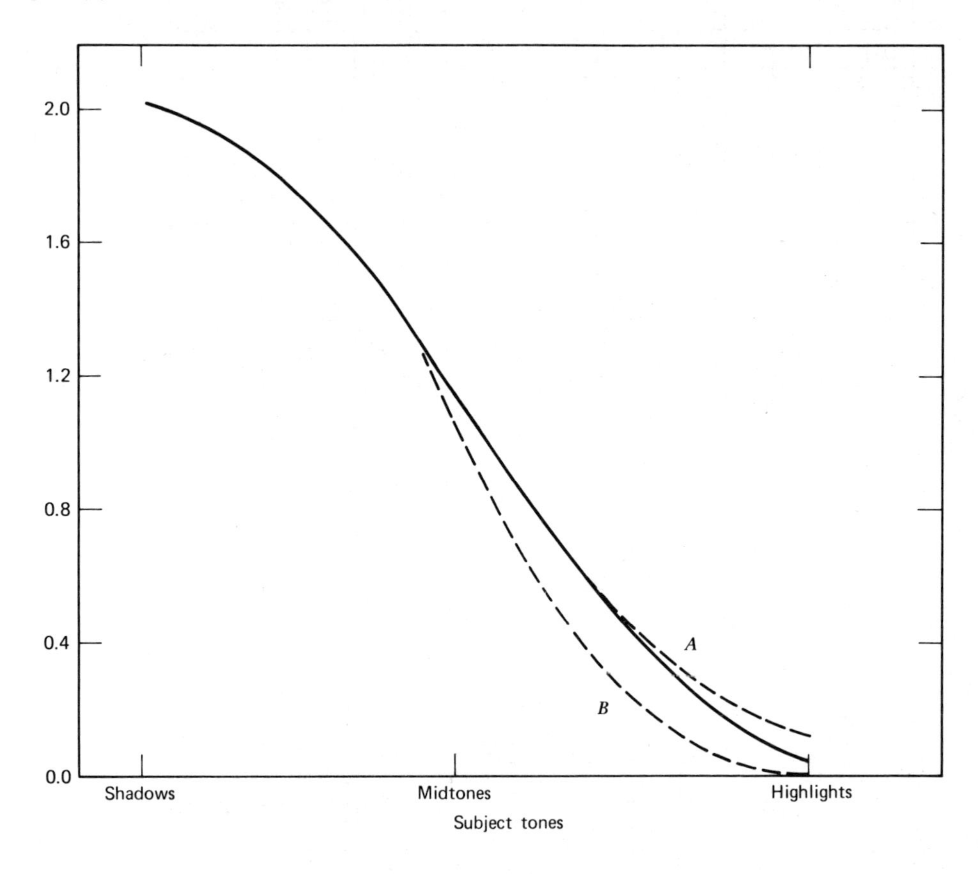

The solid line is the aim curve. Note the labeling of the subject axis.

55. Curve *A* follows the aim for the ___________ and the ___________ of the subject.
    — shadows, midtones (either order)
56. Curve *A* differs from the aim in the ___________ region.
    — highlight
57. The main difference between curve *A* and the aim is that the minimum density in the extreme highlight is about ___________ .
    — 0.12
58. Thus this highlight and the adjacent highlight tones are too ________ .
    light, dark
    — dark
59. Viewers' eyes are very sensitive to small differences in the highlight tones. Curve *A* represents a print that would appear to be dull (to lack "snap") because it would not contain any very ___________ tones.
    — light
60. For good print quality, it is important that the lightest tone be quite near to the minimum density of ___________ .
    — 0.04
61. Curve *B* shows the opposite error. The lightest tone is reproduced at a density of practically ___________ .
    — 0
62. Thus the extreme highlights are too ___________ .
    — light

63. What is worse about print *B*, however, is that the curve is nearly horizontal at the extreme right. A horizontal part of the curve means that there is practically no difference in the print tones, when there were certainly differences in the subject. In this case, the extreme __________ tones would have no detail, meaning no differences. — highlight

64. If the tone-reproduction curve is too flat (too nearly horizontal) in the highlight regions, it shows what are often called "blocked-up," or "chalky" highlights. Such an appearance in the print is especially unpleasant in portraiture. For adequate detail in the highlights (and for other tones as well) the curve must not be __________ . — flat, or horizontal

65. Curve *B* shows another print error; it is steeper than the aim for some of the __________ . Now consider the curves in the plot below. Again, the solid line is the aim for a preferred print. — midtones

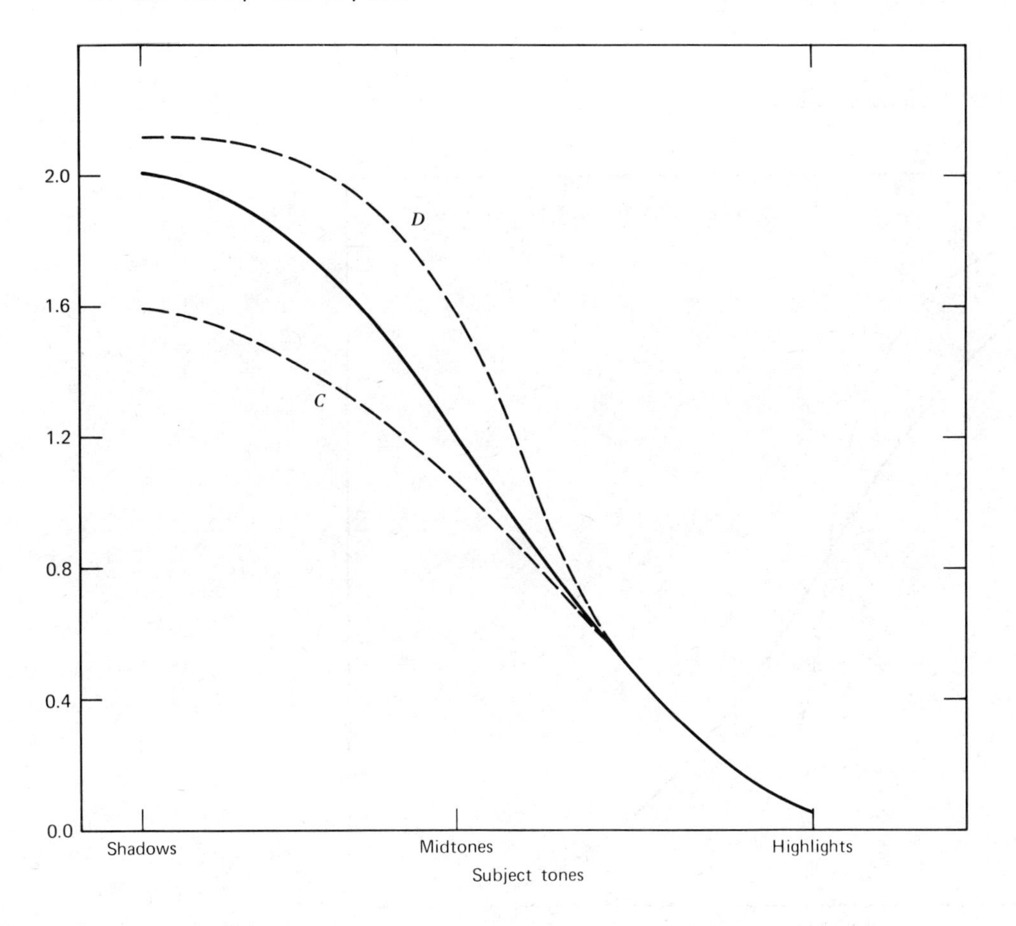

66. In the print represented by curve *C*, the highlights are correct. The largest density in this print, however, is only __________ . — 1.60

67. For good quality, the darkest shadow needs to be close to the maximum useful density, which is for a glossy paper about __________ . — 2.0

68. Thus in print C the shadows are too __________ (light, dark). In this print, the dark tones would look gray, not deep black as they should. — light

69. Also, in curve C the darker midtones are shown much less steep than the aim. A

less steep (more nearly flat) portion of the curve indicates too ________ detail (or
little, much
contrast) in these print tones. — little

70. In summary, curve C represents a print in which the shadows are too ________ and the darker midtones are too ________ . — light, flat

    This print would look unattractive because of the dull appearance of the dark tones.

71. Curve *D* shows the opposite defects. The midtones are too ________ . — contrasty
72. The shadows generally lie above the aim curve; they are too ________ . — dark
73. Also, the shadows are too ________ . — flat
74. We know that the shadows are too flat because they lie on a part of the curve that is nearly ________ . — horizontal
75. The print represented by curve *D* would have little detail in the ________ tones. — shadow or dark

As a final example, consider the following curve, again in comparison with the aim.

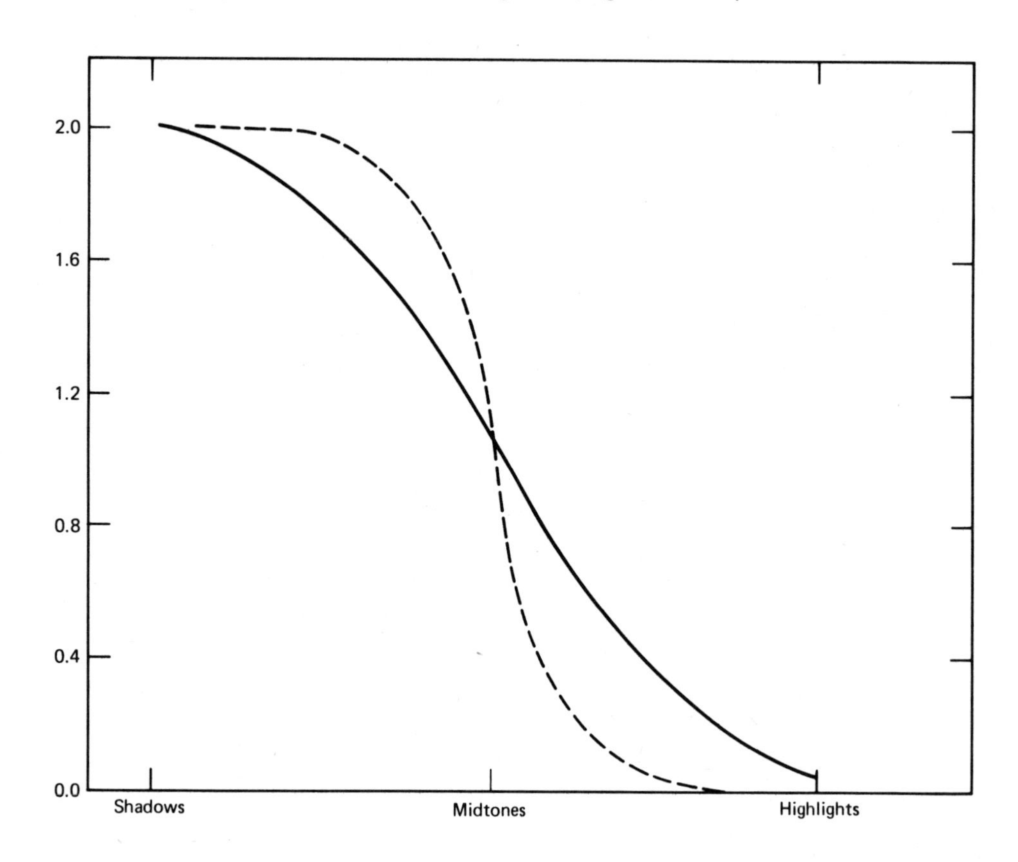

The dotted line represents a truly disastrous print.

76. The shadows are too ________ and too ________ . dark, flat (either order)

77. The midtones are much too ________ . contrasty

78. The highlights are too ________ and too ________ . light, flat (either order)

Finally, tone-reproduction plots, useful as they are, are by no means substitutes for the careful visual examination of a print by an informed viewer. Print "quality" is too subtle to be defined even by the most sophisticated measurements.

SELF-TEST ON TONE REPRODUCTION

Test your understanding of this chapter by answering the following questions. The correct answers follow.

1. The tones of a portrait subject were measured with a luminance meter. The readings were as follows: right side of face, 6; left side of face, 7; hair, 4; white blouse, 9; dark background, 3; shadow on background, 2. Mark the scale below with the values properly spaced, using a scale of 1 inch = 0.40.

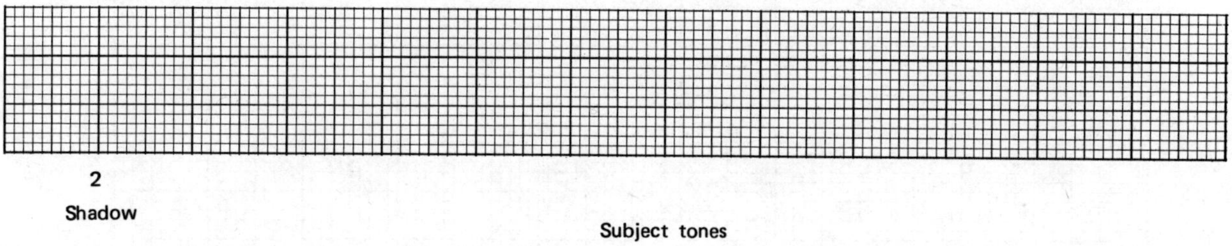

2
Shadow
Subject tones

2. In comparison with the range of a normal outdoor scene, the range of the subject in question 1 is ________________ .
flat, contrasty, near normal

3. For the same situation as in question 1, densities in the print were measured as follows: blouse, 0.00; left side of face, 0.15; right side of face, 0.32; hair, 0.84; background, 1.15; shadow, 1.52. Plot the tone-reproduction curve on the graph paper on page 158.

4. Compare the resulting curve with the aim curve shown in items 54 and 65 above:
   The highlights are ________ .
   The midtones are ________ .
   The darkest tones are ________ .

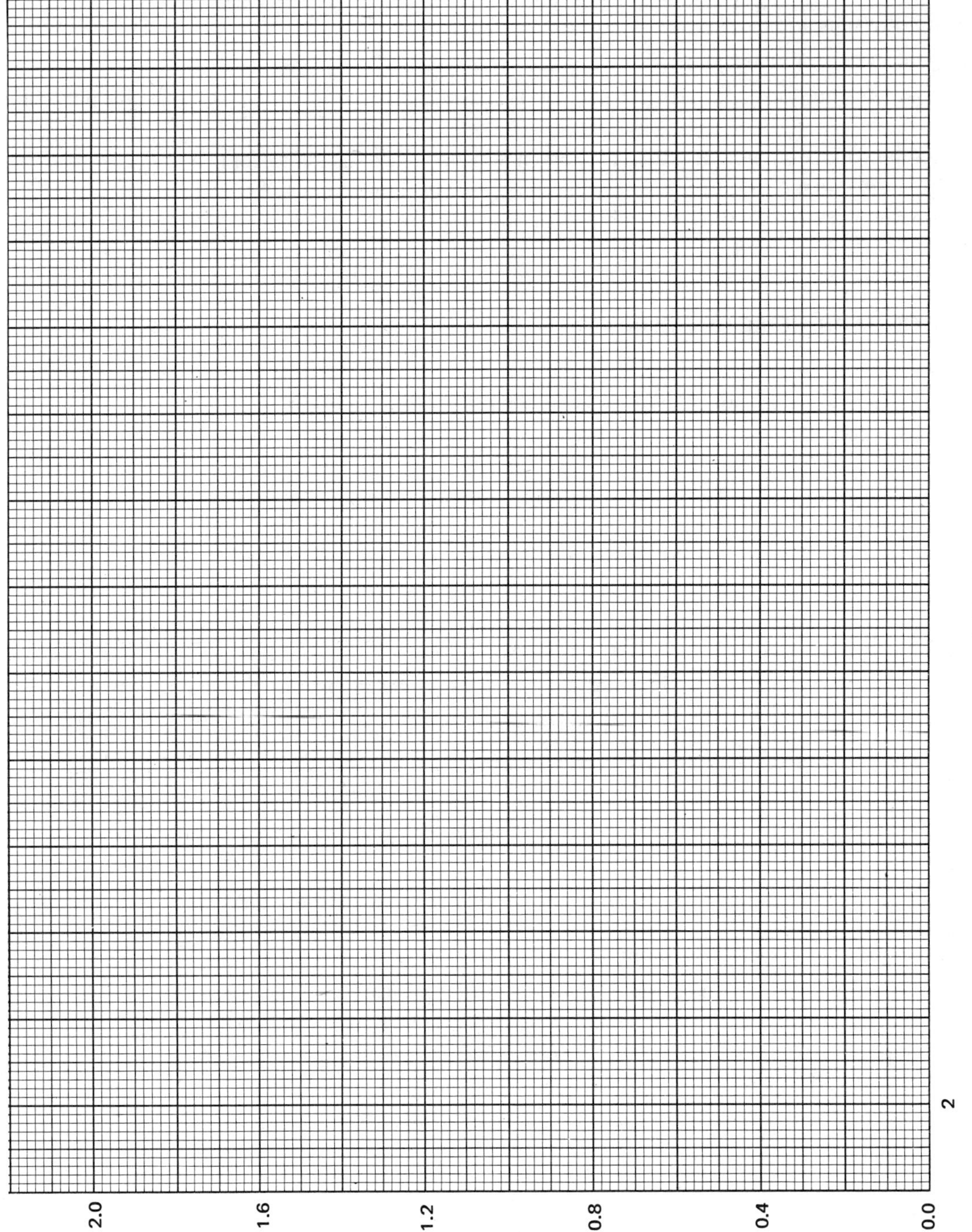
2.0
1.6
1.2
0.8
0.4
0.0
2
Subject tones

ANSWERS TO SELF-TEST ON TONE REPRODUCTION

In the parentheses after each answer you find the numbers of the items in this chapter that relate to the question and its answer.

1.

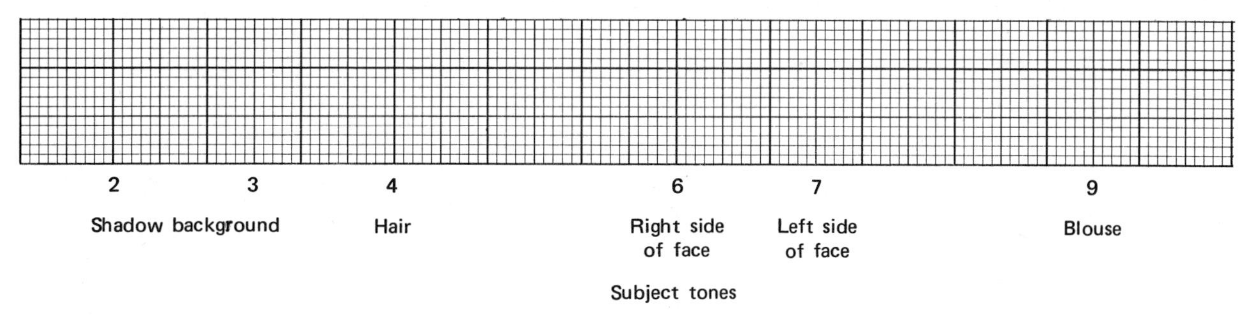

(items 9, 10, 44).

2. Near normal (items 28, 29).

3.

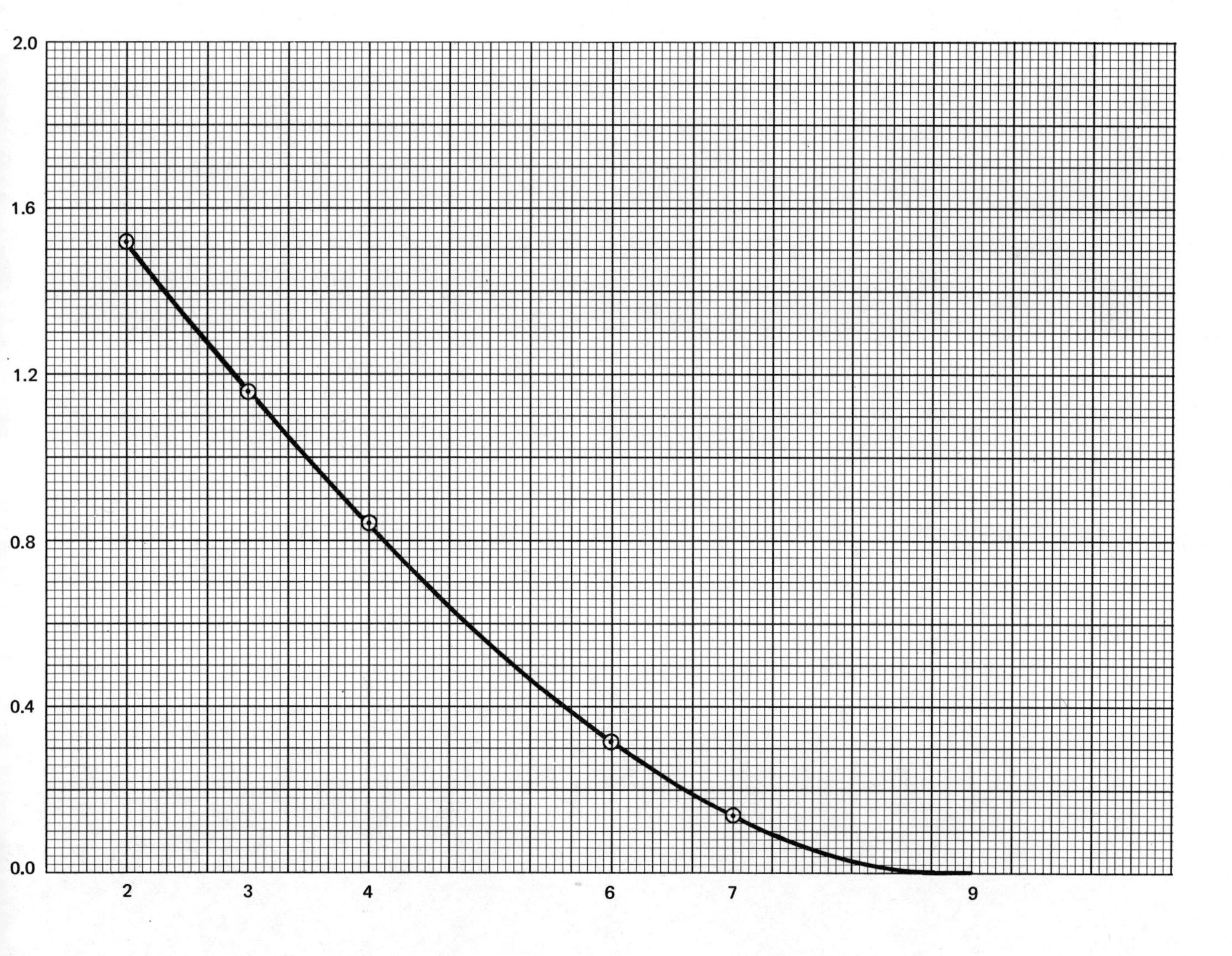

4. The highlights are too light and flat (items 61–64). The midtones are too light and flat (items 68, 69). The shadows are too light and contrasty (items 66–68, but note in the curve that the slope of the darkest tones is *greater* than that of the aim curve).

APPENDIX A
Logarithms in Photography

SECTION A1 INTRODUCTION

In photography we find many sets of numbers arranged in a series or sequence. Examples are shutter settings as marked on lenses and film-speed values. The numbers are related to each other by a common multiple, in other words, by a common factor.

For example, shutter settings on modern lenses show a series which is, in part: 1/200, 1/100, 1/50, 1/25. The times in this sequence increase by a factor of 2; that is, each value is twice the preceding value.

The standard series of film speeds is, in part: 50, 64, 80, 100, 125, 160, 200. Starting with the first number, we skip the next two, and so on, and have 50, 100, 200—again a multiplication of the values by 2. The intermediate values are related also by a factor, but smaller than 2. The factor is about 1¼.

All such number sequences that involve successive multiplication by the same factor are most easily understood and handled by the use of special numbers called "logarithms," or "logs" for short. Anyone who is working seriously with photography is using logs, consciously or not, as the following examples indicate:

1. In the zone system of Ansel Adams and Minor White, zones are values applied to the subject and to the final image. Zones are directly associated with logarithms.
2. Most modern exposure meters have scales that are related to logarithms, although the numbers themselves are not logs.

3. Every stop increase in camera exposure causes an increase in light level by a factor of 2, which is, in fact, a logarithmic change.
4. The "characteristic" curve is used in manufacturers' data sheets to show how a photographic material responds to different quantities of light. In such a plot, the numbers on both the horizontal and vertical scale are logs.
5. Both neutral density filters and color correction (CC) filters are identified with numbers that are logs.
6. The Kodak Reflection Gray Scale is commonly used for testing purposes. The numbers beside each patch of the gray scale are logarithms.
7. Our eyes, obviously important in the visual communication process, work on a nearly logarithmic basis. If we want someone to see a steady increase in the brightness of a lamp, for example, we must cause the light level to increase by steady multiplication by some number or, in other words, to increase logarithmically.

These examples show that the whole photographic process is involved with logarithms. The contents of this appendix were written to help photographers to understand logarithms and to apply them to make their photography more effective.

WHAT YOU WILL LEARN

In the six sections of this appendix you will find material that will enable you to understand the fundamentals of logarithms. If you go through these sections carefully, you will be able to:

1. Find the logarithm of any number;
2. Find the number that corresponds to any log;
3. Find the difference between the logs of two numbers when you know their ratio;
4. Find the ratio of two numbers when you know the difference between their logarithms;
5. Find the difference between any pair of logarithms;
6. Write the logarithm of a fraction in any of several different ways.

In the first section of this appendix you will learn:

1. That the logarithm of a number indicates how many tens are conmtained as multipliers in the number;
2. How to find the logarithm of any number beginning with 1 and followed by zeros, like 10,000;
3. That when a number is multiplied by 10 its logarithm is increased by 1;
4. That when a number is divided by 10 its logarithm is decreased by 1.
5. Why we use logarithms in describing the exposure of a photographic material.

DIRECTIONS. Cover about 2 inches of the right-hand margin of each of the following pages in turn with an opaque sheet of paper. Read carefully the first numbered statement. *Write* the word or phrase that you believe correctly completes the statement. Move the cover sheet down to reveal the correct answer in the margin. Continue in this way until you have finished the section. Work at a rate that is comfortable for you, and no longer than you find pleasant. Begin when you wish.

1. We often write 10 × 10 as 10 "squared," or 10 __________ . — 2
2. In the term 10^2, the number above and to the right (superscript) shows how many tens are contained as multipliers (or factors) in the equivalent number 100. Here the number of tens is __________ . — 2
3. Similarly, we can write 1000 = 10 × 10 × 10, or 10 __________ . — 3
4. In the number 1000, there are __________ tens as factors. — 3
5. In the same way, we can write 10,000 as 10 __________ . — 4
6. 10,000 is made up of __________ tens used as multipliers. — 4
7. Continuing, 100,000 is the same as 10 __________ . — 5
8. The number of tens, used as factors, in the number 100,000 is __________ . — 5
9. There are six tens in the number 1,000,000. Thus 1,000,000 = 10__________ . — 6
10. Since 10 itself contains only one ten, we write 10 = 10 __________ . — 1
11. We can now write the series:

10,	100,	1000,	10,000,	100,000,	1,000,000 thus:
↓	↓	↓	↓	↓	↓
10___	10___	10___	10___	10___	10___

1, 2, 3, 4, 5, 6

12. In the bottom row, the values above and to the right of the 10 show the number of __________ contained as factors in the original numbers. — tens
13. The number of tens contained in each number is called a *logarithm*. By this definition, the logarithm of 100 is 2, because 100 = 10 __________ . — 2
14. Similarly, because 1000 = 10^3, the log of 1000 is __________ . — 3
15. By the same definition, the log of 10,000 is __________ . — 4
16. The log of 100,000 is __________ ; the log of 1,000,000 is __________ . — 5, 6
17. For numbers beginning with 1 and otherwise containing only zeroes (like 1,000,000,000) the log is just the number of zeroes; in the case of the preceding number the log is __________ . — 9
18. By counting zeroes, the log of 10,000,000 is __________ . — 7
19. A better way of finding the log of a number (because it works for any number) is to count the number of places from the decimal point to a spot just to the *right* of the 1. For 10,000,000 you count seven places, so the log of 10,000,000 is __________ . — 7
20. A very large number begins with 1 and otherwise contains only zeroes. There are ten places from the decimal point to the right of the 1. The log of this large number is __________ . — 10
21. Every time we multiply such a number by 10, we produce another zero in the answer. Therefore the logarithm of every number increases by 1 when we multiply the original number by __________ . — 10
22. If we multiply a number by 100, we produce *two* more zeroes in the resulting answer, and the log of the original number is increased by __________ in the answer. — 2

23. Thus the rule is that multiplication by any number of tens increases the original logarithm by just the number of tens. Since 1000 contains three tens as factors, multiplying a number by 1000 would increase the original logarithm by ________ . 3

24. Conversely, *dividing* by 10 would reduce the number of zeroes in the original numeral to ________ fewer. 1

25. For this reason, dividing by 10 would reduce the logarithm of the original number by ________ . 1

26. If we were to divide a number by 100, we would reduce the logarithm by ________ . 2

27. Thus multiplication involves a(n) ________ in the logarithm of a number. increase

28. Division involves a(n) ________ in the logarithm of a number. decrease

29. In every case of multiplication or division by some number of tens, the logarithm changes by the number of ________ . tens

30. We can show the relationship between numbers and their logarithms by using two related lines, as follows:

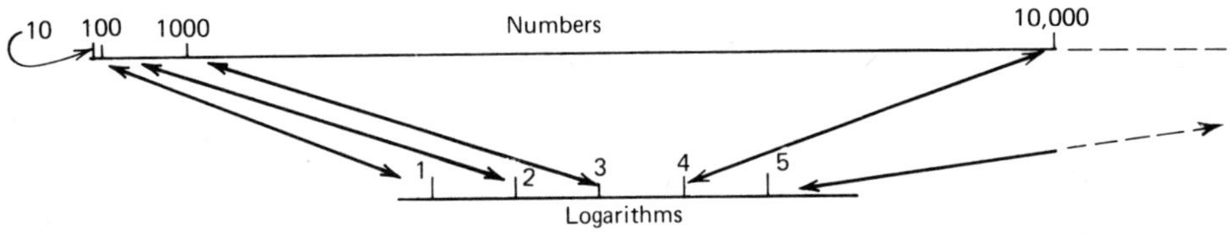

On the number line we have used a scale of 1 cm = 1000, and on the logarithm line a scale of 1 cm = 1. The next value on the number line cannot be shown on this scale, but since it corresponds to a log of 5, the number would be ________ . 100,000

31. In the preceding diagram the arrows show the important change in the scaling of values when plain numbers are represented by their logs. The scale change shows a relative *expansion* of the intervals in the logs of small numbers like 10 and 100, and a relative *compression* of the logs of ________ numbers. large

32. The reason for the scale change when we go from plain numbers to logs is that when we multiply any number by 10, the logarithm always increases by ________ . 1

33. If a small number is multiplied by 10, it increases only a little, but its log is increased by ________ . 1

34. If a very large number is multiplied by 10, it increases very greatly, but its log nevertheless ________ by only 1. increases

35. Every pair of numbers in which one is 10 times the other (ie, their *ratio* is 10-1) will have logs that differ by ________ . 1

36. Similarly, if the ratio of a pair of numbers is 100, their logs will differ by the log of 100, which is ________ . 2

37. If we have a series of numbers increasing by a common ratio (multiplication) when we change to logarithms, we would have a uniform ________ in their logs. increase

38. Conversely, to change back from a uniformly increasing set of logarithms would give us a set of plain numbers that would increase by a common ________ . ratio

39. We make use of a logarithmic scale whenever we are dealing with a set of numbers where the *ratios* of the numbers are important. We said in the introduction to this appendix that to cause a person to perceive a constant increase in brightness of a lamp we must approximately constantly multiply the output of the lamp. For this reason, we would use a scale of ________ in this situation. — logarithms

40. In photography, the important change in a light level is also indicated by a ratio. Here also we use a scale related to ________ . — logarithms

41. We use a logarithmic scale when we want to represent both very small and very large numbers together. Because of the scale transformation when we use logs, we can show both very large and very small numbers on the same line. For example, on the line below:

1 2 3 4 5 6 7 8 9 10

Scale of logarithms

the leftmost value represents the number ________ . — 10

42. On the same line, the rightmost value represents the number ________ . — 10,000,000,000

43. When we plot a graph showing the results of testing a photographic material (a *D*–log *H* curve) we usually employ a scale of logs for two reasons: (a) it is the ratio of exposures that counts and (b) we want to show the results for both very small and very ________ exposures. — large

44. Therefore on the horizontal axis of the *D*–log *H* curve you will see a scale like this:

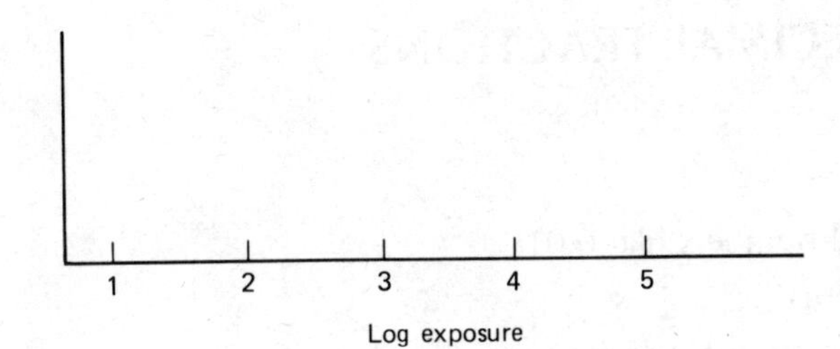

The leftmost value represents ________ units of exposure. — 10

45. On the same scale, the rightmost value represents ________ units of exposure. — 100,000

46. The equal intervals on the log-exposure axis represent, for the numbers they stand for, equal ________ of those numbers. — ratios

SELF-TEST ON LOGARITHMS (SECTION A1)

Check your understanding of this section by answering the following questions. The correct answers follow the questions.

1. How many tens, multiplied together, are contained in the number:
   (a) 100; (b) 10,000; (c) 10.
2. What is the logarithm of:
   (a) 1000; (b) 100,000; (c) 10.

3. By how much is the logarithm of a number increased if the number is multiplied by: (a) 10; (b) 100; (c) 1000.
4. By how much is the logarithm of a number decreased if the number is divided by: (a) 100; (b) 10; (c) 1000.
5. We use a scale of logarithms for two basic reasons. Give the reasons:
   (a)
   (b)

ANSWERS TO SELF-TEST ON LOGARITHMS (SECTION A1)

In the parentheses that follow each answer you find the items in this section that relate to the question and its answer.

1. (a) 2 (items 2); (b) 4 (items 5 and 6); (c) 1 (item 10).
2. (a) 3 (items 13 and 14); (b) 5 (items 13 and 20); (c) 1 (items 13 and 30).
3. (a) 1 (item 21); (b) 2 (item 22); (c) 3 (item 23).
4. (a) 2 (items 23 and 26); (b) 1 items 23 and 26); (c) 3 (items 28 and 29).
5. (a) To show both very small and very large numbers on the same scale (items 30, 41, and 42).
   (b) To show equal ratios of numbers as equal differences on a scale (items 32–40).

SECTION A2
LOGARITHMS OF SIMPLE DECIMAL FRACTIONS

In this section you will learn:

1. How to find the logarithms of fractional numbers like 0.01;
2. That all fractions have *negative* logarithms;
3. To understand the relation between the system of numbers and their logarithms.

DIRECTIONS. Cover about 2 inches of the right-hand margin of each of the following pages in turn with an opaque sheet of paper. Read carefully the first numbered statement. *Write* the word or phrase that you believe correctly completes the statement. Move the cover sheet down to reveal the correct answer in the margin. Continue in this way until you have finished the section. Work at a rate that is comfortable for you, and no longer than you find pleasant. Begin when you wish.

In Section A1 you learned about log fundamentals—the logs of numbers like 10, 100, . . ., 1,000,000. You also learned that ratios of numbers are indicated by increases in logarithms. In this section you will learn about the logarithms of numbers smaller than 10.

1. When we multiply a number by 10, its log ____________ by 1. — increases
2. Conversely, when we divide a number by 10, its log ____________ by 1. — decreases
3. Note that 1 = 10 divided by ____________ . — 10

4. Thus the log of 1 must be __________ less than the log of 10. — 1
5. Since the log of 10 is 1, the log of 1 must be __________ . — 0
6. It is logical for the log of 1 to be zero, since the log of any number starting with 1 can be found by counting the number of zeroes, and there are __________ zeroes in the numeral 1. — no
7. We can also find the log of such a number by counting the decimal places to the right of the 1. Since here the number of places is 0, the logarithm of 1 must be __________ . — 0
8. We have, thus far, this series of numbers and their logs:

Number	1	10	100	1000	...
Logarithm	0	1	2	3	...

We will now follow the same pattern and extend the series to the left.

In the *number* series, each value is one-tenth as much as its neighbor to the right. The number next to the left must be one-tenth as much as its neighbor 1, or __________ . — 0.1 or 1/10

9. Now we have:

Number	0.1	1	10	100	1000	...
Logarithm		0	1	2	3	...

In the *logarithm* series, as we go to the left each log is 1 *less* than its neighbor to the right. The next log in this series must be 1 less than 0, or __________ . — –1

10. The rule is that all numbers less than 1 have negative logs. Following this rule, and extending the series farther to the left, we have:

Number	0.01	0.1	1	10	100	1000	...
Logarithm		–1	0	1	2	3	...

The log of 0.01 must be 1 less than the log of 0.1, or 1 less than –1, which is __________ . — –2

11. All fractions have negative logs. Extending the series still farther to the left, the next member of the number series is 0.001, and its logarithm is __________ . — –3
12. We can find the log of any decimal fraction containing (besides zeros) only a 1, by counting the number of places from the decimal point to a spot just to the *right* of the 1. Thus the decimal fraction 0.0001 has a log of __________ . — –4
13. Whenever you see a negative logarithm, you know that the number is __________ than 1. — less
14. If the log of a number is –2, the minus sign means that the number is a __________ . — fraction
15. For the same logarithm (–2) the 2 indicates the number of places to the decimal point from the *right* of the 1 in the number. We write the number having a log of –2 as 0.01, putting the decimal point __________ places to the *left* of the 1. — 2
16. If the log is –3, we must put the decimal point __________ places from the right of the 1. — 3

17. Thus writing a fraction because the log is negative, the number having a log of –3 is ________ .

0.001

18. If the log is –4, the number is ________ .

0.0001

19. If we write the decimal fraction as a common fraction, the denominator shows how many tens are contained as factors in the number. Thus 0.01 = 1/100, and the denominator contains ________ tens.

2

20. The log of 0.01, or 1/100, is ________ .

–2

21. In the log of 0.01 which is –2, the negative sign shows that the number is a ________ .

fraction

22. In the log of 0.01 which is –2, the 2 stands for the number of ________ contained as factors in the denominator of the equivalent common fraction.

tens

23. The basic system of numbers and their logs is now:

Number	0.0001	0.001	0.01	0.1	1	10	100	1000	10,000
	$\frac{1}{10^4}$	$\frac{1}{10^3}$	$\frac{1}{10^2}$	$\frac{1}{10^1}$	10^0	10^1	10^2	10^3	10^4
Logarithm	–4	–3	–2	–1	0	1	2	3	4

The system is symmetrical about the *number* 1 and its *logarithm* of ________ .

0

24. Numbers less than 1, in other words, fractions, have logs with ________ signs.

negative

25. Numbers greater than 1 have logs with ________ signs.

positive

26. The numerical value of the log shows the number of ________ contained as factors in the original number.

tens

SELF-TEST ON LOGARITHMS (SECTION A2)

Check your understanding of this section by answering the following questions. The correct answers follow the questions.

1. What is the logarithm of:
   (a) 1; (b) 0.1; (c) 0.001; (d) 0.0001.
2. What is the number that has a logarithm of:
   (a) 0; (b) –1; (c) –2; (d) –4.
3. All numbers larger than 1 have logarithms with ________ signs.
4. All numbers smaller than 1 have logarithms with ________ signs.
5. (Thought question.) Zero is the smallest of all possible numbers. What, then, is the logarithm of 0?

ANSWERS TO SELF-TEST ON LOGARITHMS (SECTION A2)

In the parentheses after each answer you find the items in this section that relate to the question and its answer.

1. (a) 0 (items 5–8); (b) –1 (items 8 and 9); (c) –3 (item 11); (d) –4 (items 18 and 23).

2. (a) 1 (items 5–8); (b) 0.1 (items 9 and 10); (c) 0.01 (item 15); (d) 0.0001 (items 18 and 23).
3. Positive (items 23 and 25).
4. Negative (items 23 and 24).
5. The logarithm of 0 is the largest possible negative number, namely, $-\infty$.
   As a number becomes smaller, in the series 0.1, 0.01, 0.001, . . ., its logarithm becomes a larger and larger negative value, −1, −2, −3, . . . Zero is at the end of the series of smaller and smaller fractions, thus its log is at the end of the series of negative logarithms (item 23 extended indefinitely far to the left).

SECTION A3
CHARACTERISTICS AND MANTISSAS

In this section you will learn:

1. How to find the logarithm of any number greater than 1;
2. That multiplication of numbers means adding their logarithms;
3. That dividing numbers means subtracting their logarithms;
4. What the characteristic of a logarithm is and how to find it;
5. What the mantissa of a logarithm is and how to find it;
6. How to find numbers when you know their logarithms.

DIRECTIONS. Cover about 2 inches of the right-hand margin of each of the following pages with an opaque sheet of paper. Read carefully the first numbered statement. *Write* the word or phrase that you believe correctly completes the statement. Move the cover sheet down to reveal the correct answer in the margin. Continue in this way until you have finished the section. Work at a rate that is comfortable for you, and no longer than you find pleasant. Begin when you wish.

In Sections A1 and A2 you learned how to find the log of any number, large or small, that contains (aside from zeros) only the digit 1. In this section you will learn about the logs of other numbers.

1. We know that the log of 1 is ________ and the log of 10 is ________. — 0, 1
2. Now, 2 lies between 1 and 10 on the number line. Thus its log must lie between the logs of 1 and 10, that is, between ________ and ________. — 0 and 1
3. The logs of numbers like 2 are found from a table, such as that at the end of Section A4. For now, take it that from such a table the log of 2 is approximately 0.30. Knowing this, we can find the logs of other numbers, namely, those that contain 2 as a factor. Just as multiplying a number by 10 increases the log by 1, so multiplying a number by 2 increases the log by the log of 2, or by ________. — 0.30
4. Since 4 is 2×2, the log of 4 must be greater than the log of 2 by ________. — 0.30
5. The log of 4 is 0.60, because $4 = 2 \times 2$ and the log of 2 is 0.30. Similarly 8 is 2×4, and the log of 8 must be ________ more than the log of 4. — 0.30
6. Thus from just the log of 2 we have this partial table:

Number	1	2	4	8
Logarithm	0.00	0.30	0.60	0.90

On the number line as we go to the right, each value is multiplied by ___________ . — 2

7. In the same table, on the logarithm line, as we go to the right, each value is increased by ___________ . — 0.30
8. We could extend the log table indefinitely to the right, following the same rule: multiplying successively by 2 would steadily increase the log by 0.30. Thus if the log of 8 is 0.90, the log of 16 (2 times 8) is ___________ and the log of 32 is ___________ . — 1.20, 1.50
9. If we refer to a complete table of logs, we find that it gives the log of 3 as approximately 0.48. Since 6 is 2 times three, the log of 6 must be 0.30 more than the log of 3, or ___________ . — 0.78
10. So far we have this partial list:

Number	1	2	3	4	6	8	10
Logarithm	0.00	0.30	0.48	0.60	0.78	0.90	1.00

We can insert the log for 9, because $9 = 3 \times 3$. Adding the log of 3 to itself (or more simply doubling the log of 3) we find that the log of 9 is ___________ . — 0.96

11. We can insert the log for 5 by this reasoning: 5 is 10 divided by 2, and the log of 5 must be *less* than the log of 10 by 0.30, the log of 2. Thus the log of 5 is ___________ . — 0.70
12. To complete the list of the logs of whole numbers from 1 to 10, we refer to a standard table to find that the log of 7 is approximately 0.84. Thus we have:

Number	1	2	3	4	5	6	7	8	9	10
Logarithm	0.00	0.30	0.48	0.60	0.70	0.78	0.84	0.90	0.96	1.00

From this table, we see that every *number* that is twice another *number* has a logarithm that is ___________ more. — 0.30

13. In general, the table shows that the log of a product is the sum of the ___________ of the multiplied numbers. — logs
14. Conversely, in the same table we see that when a *number* is *divided* by another number, the log of the first number is *decreased* by the log of the second number. Since 8 divided by 4 is 2, the log of 2 is ___________ less than the log of 8. — 0.60
15. We have shown by example that when numbers are multiplied, their logs are ___________ . — added
16. We have also shown by example that when we divide one number by a second, we ___________ the log of the second from that of the first. — subtract
17. The addition of logs is the equivalent of the ___________ of numbers. — multiplication
18. The *subtraction* of a log is the equivalent of ___________ . — division

For the following items refer to the table in item 12 above.

19. By application of the rules for multiplying and dividing numbers (adding and subtracting logs) we can find the logs of many numbers from the limited table above. For example, since $15 = 3 \times 5$, the log of 15 must be the ________ of the logs for 3 and 5, that is, ________ .

sum,
1.18

20. Similarly, since $35 = 5 \times 7$, the log of 35 must be the sum of the logs of 5 and 7, or ________ .

1.54

21. The log of 64 (which is 8×8) is ________ .

1.80

22. If we apply the rule for division of numbers by the subtraction of logs: $2.5 = 5$ divided by 2. Therefore the log of 2.5 must be the log of 5 less the log of 2, or 0.70 less 0.30, or 0.40. The log of 3.5 can be found in the same way, since $3.5 = 7$ divided by 2, so the log of 3.5 is ________ .

0.54

23. Using the rule for division, the log of 4.5 (which is 9 divided by 2) is ________ .

0.66

24. A special case involves the multiplication of a number by 10, or 100, and so on. For example, the log of 20 (which is 2×10) is just more than the log of 2 by the log of 10, or ________ .

1.30

25. Also, the log of 30 is 1 more than the log of 3, that is, ________ .

1.48

26. The log of 50 is then ________ .

1.70

27. From the preceding examples, you see that every number like 20, 30, 40, 50, or 90 will have a logarithm beginning with ________ , because every one of these numbers is some small value multiplied by ________ .

1, 10

28. On the other hand, 200 is 2×100, and the log of 200 must be the log of 2 increased by the log of 100, or 0.30 + ________ .

2

29. Thus the log of 200 is 2.30. The log of 400 must be ________ .

2.6

30. Every number like 200, 300, 400, . . ., 900 will have a log beginning with ________ , because every such number is some small value multiplied by ________ .

2,
100

31. By the same logic, every number in the thousands, like 4000, will have a log beginning with ________ .

3

32. Every number in the ten thousands, like 80,000, will have a log beginning with ________ .

4

33. We call the whole-number part of the log the *characteristic*. The characteristic is found for numbers like 300,000 just as for 100,000: count the decimal places found to the *right* of the first digit. For 300,000, the count is ________ .

5

35. The log of 300,000 has the characteristic 5. The whole log is ________ .

5.48

36. The characteristic of the log of 7000 is ________ .

3

37. The characteristic of the log of 6,000,000 is ________ .

6

38. The characteristic of the log of 8,500,000 is also ________ .

6

39. The rest of the logarithm of a number, the decimal part of the log, is found in a table, like that in item 12 but more extensive in general. The decimal part of the log is called the *mantissa*. The value of the mantissa is fixed by the digits in the number, regardless of the position of the decimal point. Thus for all numbers like 5, 50, 500, 5000, . . . the mantissa is identical, and is found in the table you have been using to be ________ .

0.70

40. For the set of numbers 3, 30, 300, 3000, . . . the mantissa would be 0.48, and only the whole-number part of the log, the ________ would change.

characteristic

41. The characteristics of the logs of the numbers 3, 30, 300 in order would be: _______, _______, _______.	0, 1, 2
42. For the number 70,000, the characteristic of the log is _______, from the table in item 12 the mantissa is _______, and the entire log is _______.	4, 0.84, 4.84
43. For the number 800,000, the characteristic is _______, the mantissa is _______, and the entire log is _______.	5, 0.90, 5.90
44. The entire log of 60 is _______.	1.78
45. The log of 5,000,000 is _______.	6.70
46. If the logarithm is given and the number is needed, it is found by a reverse process. The number corresponding to a given log is called the *antilogarithm,* or antilog for short. If the log is 2.78, the characteristic determines where to put the decimal point. Here the characteristic is 2, so there must be two places from the digit in the antilog to the decimal point. The mantissa determines the digit in the antilog. Here the mantissa is 0.78, which corresponds to the digit _______ on the number line of the table.	6
47. Placing the decimal point two places to the right of the 6 makes the antilogarithm (the number) _______.	600
48. If the given log is 2.96, the decimal point in the antilog is again determined by the characteristic, which is here _______.	2
49. The mantissa in the log 2.96 is _______.	0.96
50. From the table, the digit corresponding to the mantissa (0.96) is _______.	9
51. Using the characteristic (2) to put the decimal point in the right place, the antilog of 2.96 is _______.	900
52. Following the same procedure, the antilog of 1.48 is _______.	30
53. Similarly, the antilog of 6.60 is _______.	4,000,000
54. You have seen how to find the log of any number greater than 1 if it has only one digit (other than zeros). You can find the logs of numbers having more than one digit from a more nearly complete table. In most photographic situations, we rarely need to work with more than two digits in the number; at the end of Section A4 on page 178 is a table giving logs of numbers that contain two digits. Use this table for the following items. In the expanded table, to find the log of any two-digit number, find the digits of the number in the *number* column, and read the *mantissa* of the log to the right. Find the characteristic of the log as before, by counting the number of places from the right of the leftmost digit to the decimal point in the number. For example, to find the log of 650, the characteristic is _______.	2
55. Opposite 6.5 on the number line you find the mantissa, to the right in the table. The mantissa is _______.	0.813
56. Put the characteristic before the mantissa. You have the entire log of 650. The log is _______.	2.813
57. For the number 3.3, the characteristic is _______.	0
58. For the number 3.3, the mantissa is _______.	0.518
60. The entire log of the number 19,000 is _______.	4.278
61. To find the antilog of a number using the expanded table, find in the table the mantissa nearest the given value. Read in the *number* column the digits in the	

answer. Place the decimal point using the characteristic of the log as before. For example, if the given log is 2.435, find the closest value of the mantissa in the table. It is 0.431. To the left of this mantissa read the digits of the answer, which are __________ .

2.7

63. Since the characteristic is 2, you need to move the decimal point two places to the right of the lefthand digit, and thus the antilogarithm is __________ .

270

64. By the same process, the antilogarithm of 1.912 is __________ .

82

65. The antilogarithm of 4.650 is __________ .

45,000

66. The antilogarithm of 0.245 is __________ .

1.8

SELF-TEST ON LOGARITHMS (SECTION A3)

Check your understanding of this section by answering the following questions. The correct answers follow the questions.

1. The logarithm of 2 is 0.30. What is the log of each of the following numbers: (a) 4 (which is 2 × 2); (b) 8 (which is 2 × 2 × 2); (c) 16.
2. The logarithm of 3 is 0.48. What is the logarithm of 9 (which is 3 × 3)?
3. The logs of two numbers are 0.48 and 0.70. What is the logarithm of their product? What is the logarithm of their quotient when the larger number is divided by the smaller?
4. What is the characteristic (whole-number part of the log) of the logarithms of: (a) 70; (b) 90; (c) 400; (d) 6000?
5. What is the mantissa (decimal part of the logarithm) of the logs of the numbers: (a) 7; (b) 70; (c) 700.
6. What is the antilog (the number having the given log) of the following logarithms: (a) 1.78; (b) 2.96.
7. Use the appendix table of logarithms to find the logs of: (a) 650; (b) 780; (c) 5,400.
8. Use the same table to find the antilogs of: (a) 2.435; (b) 1.555; (c) 3.680.

ANSWERS TO SELF-TEST ON LOGARITHMS (SECTION A3)

In the parentheses after each answer you find the items in this section that relate to the question and its answer.

1. (a) 0.60 (items 3 and 4); (b) 0.90 (items 3 and 5); (c) 1.20 (item 8).
2. 0.96 (item 10).
3. 1.18 (items 13 and 19); 0.22 (item 22).
4. (a) 1 (items 24–27, 33, 34); (b) 1 (items 24–27, 33, 34); (c) 2 (items 28–30, 34); (d) 3 (items 34 and 35).
5. (a) 0.84 (item 12); (b) 0.84 (items 39 and 42); (c) 0.84 (items 39 and 42).

6. (a) 60 (items 46 and 47); (b) 900 (items 50–52).
7. (a) 2.813 (items 54–60); (b) 2.892 (items 54–60); (c) 3.732 (items 54–60).
8. (a) 270 (items 61–66); (b) 36 (items 61–66); (c) 4800 (items 61–66).

SECTION A4
LOGARITHMS OF FRACTIONS

We now continue the basic treatment of logarithms and deal with the logs of numbers less than 1, that is, decimal fractions.

In this section you will learn:

1. How to find the logarithm of any number less than 1, that is, any fraction;
2. That the logarithms of fractions always have negative characteristics;
3. The different ways of writing the logarithm of a fraction;
4. How to find a fractional number if you know its logarithm.

DIRECTIONS. Cover about 2 inches of the right-hand margin of each of the following pages in turn with an opaque sheet of paper. Read carefully the first numbered statement. *Write* the word or phrase that you believe correctly completes the statement. Move the cover sheet down to reveal the correct answer in the margin. Continue in this way until you have finished the section. Work at a rate that is comfortable for you, and no longer than you find pleasant. Begin when you wish.

1. You recall that the numbers in the series 0.1, 0.01, 0.001, . . . have *negative* logs, and that the value of the log is found as usual by counting decimal places. Thus the log of 0.001 is ____________ . —— –3
2. Similarly, the log of 0.00001 is ____________ . —— –5
3. Consider the *number* 0.2. It is two-tenths, that is, 2 × 0.1. We follow the rule that in multiplying numbers we add their logs. Here we would add the log of 2 to the log of 0.1. From the table of logs, we find that the log of 2 is ____________, and we know that the log of 0.1 is ____________ . —— 0.3, –1
4. Thus the computation we would carry out in finding the log of 0.2 would be: add 0.30 (the log of ____________) to –1 (the log of ____________). —— 2, 0.1
5. Therefore the calculation for the log of 0.2 would be 0.30 + (–1), or simply 0.30 –1. For the log of the number 0.50, (5/10, ie, 5 × 0.1) we follow the same procedure: add the log of 5 (from the table, ____________) to the log of 0.1, which is still –1. —— 0.70
6. Therefore, we show the calculation for the log of 0.50 as 0.70 –1. Now, for the number 0.02 (two *hundredths,* ie, 2 × 0.01), the calculation would read ____________ – ____________ . —— 0.30 –2
7. For the number 0.006, the log is ____________ – ____________ —— 0.78 –3
8. In the logs of decimal fractions, just like those for large numbers, the log contains the mantissa (found from the table) and the characteristic, a negative value, found from the location of the ____________ ____________ . (two words) —— decimal point
9. We could, but usually do not, finish the calculation indicated by the two-part

logarithm. Therefore to find the log of 0.07, find the mantissa in the table at the end of Section A4. It is 0.84. Find the characteristic by counting the number of places from the decimal point to the right of the digit, and put a minus sign in front of the characteristic. The complete log is ________ (mantissa – characteristic). — 0.84 –2

10. By leaving the indicated subtraction unfinished, we can use the same table for the logs of fractions as well as for large numbers. You just need to remember that fractions have logs with ____ (sign) characteristics. — negative

11. To find the log of 0.0095, find the mantissa for the digits 9.5 in the table at the end of this section. The mantissa is ________ . — 0.978

12. Count the places in the number from the decimal point to the right of the left-hand digit (the 9). The characteristic is ________ . — –3

13. The entire log of the number 0.0095 is therefore 0.978 –3. Following the same procedure, the log of 0.00046 is ________ (mantissa – characteristic). — 0.663 –4

14. By the same procedure, the log of 0.57 is ________ . — 0.756 –1

15. The log of 0.0000041 is ________ . — 0.613 –6

16. The log of 0.00033 is ________ . — 0.518 –4

17. The way we have written the logs of fractions is direct and reasonable. It is conventional, however, to write the logs of fractions differently. A common way (especially in photography) is to write the characteristic *in front of* the mantissa, and to put the minus sign *over* the characteristic. For example, we have written the log of 0.02 as 0.30 –2. Following this new method, we write the log as $\bar{2}.30$. Similarly, the log of 0.006 is 0.78 –3. Placing the characteristic in front of the mantissa and putting the minus sign over the characteristic, we write 0.78 –3 as ________ . — $\bar{3}.78$

18. We read a log like $\bar{3}.78$ as "bar-three point seven eight." You must remember that in this form only the *characteristic* is negative. The mantissa is positive. In this "bar" notation, the log of the number 0.46, which we previously wrote as 0.663 –1, we now write as ________ . — $\bar{1}.663$

19. When we use the bar notation, we find the log of a number just as before—look up the mantissa in the table, and find the characteristic by counting decimal places. Write the characteristic *in front of* the mantissa, and put the minus sign *over* the characteristic (indicating that this is a special way of writing *two* numbers, one negative and one positive). Thus in the bar notation, the log of 0.054 is ________ . — $\bar{2}.732$

20. In bar notation the log of 0.00061 is ________ . — $\bar{4}.785$

21. In bar notation the log of 0.079 is ________ . — $\bar{2}.898$

22. In bar notation the log of 0.000,000,82 is ________ . — $\bar{7}.914$

23. Another way of writing the log of a fraction is to change the characteristic *only* into an equivalent positive number less 10. Thus –1 is 9 –10; –2 is 8 –10; –3 is 7 –10, and so on. In each case, if we carried out the subtraction we would obtain the negative characteristic with which we started. When we use this form, the mantissa follows the positive whole number. For example, the log of 0.5 is 0.70 –1. In this "minus-ten" notation, we think that 9 –10 is equal to –1, and we write the mantissa (0.70) after the 9, thus 9.70 –10. In the first form, the log of 0.05 is 0.70 –2. In the "minus-ten" notation it is ________ –10. — 8.70

24. The log of 0.007 is initially 0.84 –3. In the "minus-ten" form it is ________ .

    7.84 –10

25. If we want the log of 0.0075, we know that the characteristic is ________ .

    –3

26. For the log of 0.0075, we find the mantissa from the table to be ________ .

    0.875

27. In the first form the log of 0.0075 is ________________ .
    mantissa – characteristic

    0.875 –3

28. In the bar notation, the log of 0.0075 is ________ .

    $\bar{3}.875$

29. In the minus-ten form, the log of 0.0075 is ________ .

    7.875 –10

30. The log of 0.00094 is in the three notations:

    ____
    mantissa – characteristic

    0.973 –4

    bar notation: ________ .

    $\bar{4}.973$

    minus-ten notation: ________ .

    6.973 –10

31. In each of these forms of the log of a fraction, the mantissa *never* changes. The only change is in the manner of writing the characteristic. The log of 0.12 is:

    ____
    mantissa – characteristic

    0.080 –1

    bar notation: ________ .

    $\bar{1}.080$

    minus-ten notation: ________ .

    9.080 –10

32. Antilogs of logarithms with negative characteristics are found just like those with positive characteristics. For example, if the log is 0.929 –2, we look up the mantissa (0.929) in the table, and find the digits 8.5. The characteristic (–2) indicates that the number (the antilog) is a fraction, and we therefore move the decimal point ________ places to the left.

    2

33. The antilog of 0.929 –2 is therefore ________ .

    0.085

34. By the same process, the antilog of 0.826 –1 is ________ .

    0.67

35. If the log is given in the bar notation, such as $\bar{3}.544$, we again look up the mantissa in the table and find the digits 3.5. In this notation, the characteristic is ________ .

    –3

36. The characteristic of –3 indicates that the decimal point must be moved 3 places to the left, and thus the antilog of $\bar{3}.544$ is ________ .

    0.0035

37. The antilog of $\bar{2}.230$ is ________ .

    0.017

38. Finding the nearest mantissa in the table, the antilog of $\bar{1}.370$ is ________ .

    0.23

39. In the same way, the antilog of $\bar{4}.688$ is ________ .

    0.00049

40. If the log is given in the minus-ten notation, we look up the mantissa as before and find the digits in the antilog. We combine the whole-number parts of the log to find the number of places the decimal point must be moved. For example, if the log is 8.857 –10, the mantissa gives us the digits ________.

    7.2

41. For the log 8.857 –10, the digits are 7.2. We combine the 8 and the –10 to get –2, indicating that the decimal point must be moved two places to the left. Thus the resulting antilog is ________ .

    0.072

42. If the log is 7.785 –10, the digits in the antilog are, from the table, ________ .

    6.1

43. Since 7 –10 is –3, the decimal point must be moved three places to the left, and the antilog of 7.785 –10 is thus ________ .

    0.0061

44. The antilog of 8.890 –10 is ________ .

    0.078

45. The antilog of 9.845 –10 is ________ .

    0.70

SELF-TEST ON LOGARITHMS (SECTION A4)

Check your understanding of this section by answering the following questions. The correct answers follow the questions.

1. For the number 0.2 the mantissa of its logarithm is ___________ .
   For the same number the characteristic of its logarithm is ___________ .
2. For the number 0.02 the mantissa of its log is ___________ .
   The characteristic of its log is ___________ .
3. The logarithm of 0.0095 is ___________________ .
   mantissa – characteristic
4. The logarithm of 0.041 is ___________________ .
   mantissa – characteristic
5. In bar notation, the logarithm of 0.041 is ___________ .
6. In "minus-ten" notation, the logarithm of 0.041 is ___________ .
7. The logarithm of 0.63 is ___________________ .
   mantissa – characteristic
8. In bar notation the logarithm of 0.63 is ___________ .
9. In "minus-ten" notation the logarithm of 0.63 is ___________ .
10. The antilog of 0.929 –2 is ___________ .
11. The antilog of 0.556 –1 is ___________ .
12. The antilog of $\bar{2}.230$ is ___________ .
13. The antilog of $\bar{1}.700$ is ___________ .
14. The antilog of 8.579 –10 is ___________ .

ANSWERS TO SELF-TEST ON LOGARITHMS (SECTION A4)

In the parentheses after each answer you find the items in this section that relate to the question and its answer.

1. (a) 0.30 (items 3–4); (b) –1 (items 3–6).
2. (a) 0.30 (items 3–4); (b) –2 (items 3, 4, 6).
3. 0.978 –3 (items 11–12).
4. 0.613 –2 (items 9, 15).
5. $\bar{2}.613$ (items 17–19).
6. 8.613 –10 (item 23).
7. 0.799 –1 (items 3–4).
8. $\bar{1}.799$ (items 17–19).
9. 9.799 –10 (item 23).
10. 0.085 (items 32, 33).
11. 0.36 (items 32, 33).
12. 0.017 (items 35–37).
13. 0.50 (items 35–37).
14. 0.038 (items 42–43).

ABBREVIATED TABLE OF LOGARITHMS

Number	Logarithm
1.0	0.000
1.1	0.042
1.2	0.080
1.26	0.100
1.3	0.114
1.4	0.147
1.414	0.150
1.5	0.175
1.6	0.204
1.7	0.230
1.8	0.255
1.9	0.278
2.0	0.301
2.1	0.322
2.2	0.342
2.3	0.361
2.4	0.390
2.5	0.398
2.6	0.415
2.7	0.431
2.8	0.447
2.9	0.462
3.0	0.477
3.1	0.491
3.2	0.505
3.3	0.518
3.4	0.532
3.5	0.544
3.6	0.556
3.7	0.568
3.8	0.580
3.9	0.591
4.0	0.602
4.1	0.613
4.2	0.623
4.3	0.634
4.4	0.644
4.5	0.653
4.6	0.663
4.7	0.672
4.8	0.681
4.9	0.690

Number	Logarithm
5.0	0.699
5.1	0.708
5.2	0.716
5.3	0.724
5.4	0.732
5.5	0.740
5.6	0.748
5.7	0.756
5.8	0.763
5.9	0.771
6.0	0.778
6.1	0.785
6.2	0.792
6.3	0.799
6.4	0.806
6.5	0.813
6.6	0.819
6.7	0.826
6.8	0.832
6.9	0.839
7.0	0.845
7.1	0.851
7.2	0.857
7.3	0.863
7.4	0.869
7.5	0.875
7.6	0.881
7.7	0.886
7.8	0.892
7.9	0.898
8.0	0.903
8.1	0.908
8.2	0.914
8.3	0.919
8.4	0.924
8.5	0.929
8.6	0.934
8.7	0.940
8.8	0.944
8.9	0.949
9.0	0.954
9.1	0.959
9.2	0.964
9.3	0.968
9.4	0.973

Number	Logarithm
9.5	0.978
9.6	0.982
9.7	0.987
9.8	0.991
9.9	0.996
10.0	1.000
0.0001	−4
0.001	−3
0.01	−2
0.1	−1
1	0
10	1
100	2
1000	3
10,000	4
100,000	5
2	0.301
20	1.301
200	2.301
2000	3.301
0.2	0.301-1 or 9.301-10 or $\bar{1}.301$
0.02	0.301-2 or 8.301-10 or $\bar{2}.301$
30	1.477
0.3	0.477-1 or 9.477-10 or $\bar{1}.477$

SECTION A5
THE DIFFERENCE BETWEEN TWO LOGARITHMS

In Sections A1 and A3 you saw that if a number is multiplied by some value, the log of the number is increased by the log of the value. For example, if a number is multiplied by 10, the log of the original number is increased by 1 (the log of the factor 10).

Conversely, if we find the *difference* in two logs, that difference is the log of the *ratio* of the two numbers belonging to those logs. In photographic applications the ratio of numbers such as film speeds and shutter settings can be related to stops, and thus such a ratio has great usefulness. For the same reason, to be able to find the difference in a pair of logs is often necessary.

WHAT YOU WILL LEARN

If you work carefully with the section that follows, you will be able to:

1. Find the difference between any pair of logarithms;
2. From the difference between the logs, find the ratio of the two numbers represented by the logs.

DIRECTIONS. Cover about 2 inches of the right-hand margin of each of the following pages in turn with an opaque sheet of paper. Read carefully the first numbered statement. *Write* the word or phrase that you believe correctly completes the statement. Move the cover sheet down to reveal the correct answer in the margin. Continue in this way until you have finished the section. Work at a rate that is comfortable for you, and no longer than you find pleasant. Begin when you wish.

1. If the logs of two numbers are 1.60 and 1.30, the difference between these logs is ________.
   0.30
2. The antilog of the difference (0.30) is the ratio of the two *numbers*. The antilog of 0.30 is ________ .
   2
3. This result (2) means that the larger of the two numbers is 2 times the smaller. See that the same result appears if you find the antilogs of the given logs (1.60 and 1.30), which are 40 and 20. Thus the larger number is ________ times the smaller.
   2
4. Similarly, if the logs of two numbers are 2.60 and 2.30 the difference in the logs is again ________ .
   0.30
5. Because the log difference is 0.30 we know that the larger number (of the two having these logs) is ________ times the smaller number.
   2
6. You would get the same result, regardless of the size of the logs (or their antilogs); if two logs differ by 0.30, the numbers they stand for would be related by a ratio of ________ to 1.
   2
7. By the same process, if two numbers have logs that differ by 0.60, the numbers themselves are related by a factor of ________ (the antilog of 0.60).
   4
8. If two logs differ by 0.90, the numbers are related by a ________ (multiplier) of ________ .
   factor, 8

9. The logs of two numbers are 2.78 and 2.00. Without looking up the numbers having these logs, we see that the difference in the logs is ______ . | 0.78
10. If the difference in the logs of two numbers is 0.78, the ratio of the numbers must be ______ to 1. | 6
11. The answer just found, 6, means that one number is six times the other. If two logs are 3.45 and 1.85, the difference in these logs is ______ . | 1.60
12. The ratio of the two numbers is the antilog of 1.60, which is ______ . | 40
13. Thus the larger of the two numbers is 40 times the smaller. If the logs of two numbers are 1.92 and 0.32, in this case also the larger number is ______ times the smaller. | 40
14. If the logs of two numbers are 1.78 and 1.18, the larger of the numbers is ______ times the smaller. | 4
15. If two logs are 3.00 and 2.52, the larger of the two numbers represented by these logs is ______ times the smaller number. | 3
16. If two logs are 3.00 and 1.52, the larger of the two numbers having these logs is ______ times the smaller. | 30
17. Two logs are 1.00 and 0.17. The ratio of the numbers having these logs is ______ to 1. | 6.8
18. The rule that a log difference implies the ratio of the corresponding plain numbers applies to all logs and all numbers. Subtraction of logs must be done carefully for the logs of fractions, that is, for logs having negative characteristics. If the two logs are 0.60 –1 and 0.30 –1, we set up the subtraction thus: $\begin{array}{r} 0.60 - 1 \\ \underline{0.30 - 1} \end{array}$ with the *larger* log (0.60 –1) on top. We need to subtract both the characteristics and the mantissas. Since the characteristics are equal, when we subtract them we get ______ . | 0
19. Now, we subtract the mantissas: $\begin{array}{r} 0.60 \\ \underline{0.30} \end{array}$ and get ______ . | 0.30
20. The antilog of the difference in the logs (0.30) is the ratio of the numbers represented by the logs we started with. The ratio is ______ to 1. | 2
21. If the given logs are 0.78 –2 and 0.30 –2, we set up the subtraction the same way: $\begin{array}{r} 0.78 - 2 \\ \underline{0.30 - 2} \end{array}$. Again, since the characteristics are equal, their difference is | 0
22. The difference in the mantissas is ______ . | 0.48
23. Since we just found that the log difference is 0.48, the numbers must have a ratio of ______ to 1. | 3
24. The ratio of two numbers having logs of 0.40 –2 and 0.10 –2 is ______ to 1. | 2
25. If the logs of two numbers are 0.90 –3 and 0.30 –3, their ratio is ______ to 1. | 4
26. When the characteristics of two logs are identical, the difference in the characteristics is always ______ . | 0
27. When the characteristics are the same, we need therefore only subtract the ______ of the logs. | mantissas
28. When the characteristics are different and the mantissas are the same, we then need only to subtract the characteristics, since the difference in the mantissas will

be zero. For example if the logs are 0.78 –1 and 0.78 –2, we set up the subtraction thus: 0.78 –1, the *larger* value on *top* of the smaller (0.78 –1 is *larger* because it
<u>0.78 –2</u>

represents a number in the *tenths,* whereas 0.78 –2 represents a number in the *hundredths*). To subtract the characteristic of –2 from the characteristic of –1, change the sign of the –2 to +2, and combine with the –1 to get ______ . (+)1

29. Thus if the characteristics are –2 and –1, the *difference* in the logs is 1, indicating that the larger number is (from the antilog of 1) ______ times the smaller. 10

30. By the same process, if the two logs are 0.60 –2 and 0.60 –1, we write the subtraction: 0.60 –1, placing the *larger* value on top. Again see that the mantissas
<u>0.60 –2</u>

are identical and thus have a difference of 0. Subtract –2 from –1 by changing the sign of the –2 to +2, combine with –1 and get ______ . (+) 1

32. If the two logs are 0.96 –1 and 0.96 –3, their difference is found by subtracting the characteristics only, to get ______. 2

33. That the difference in the logs is 2 means that the numbers represented by these logs differ by a factor of ______. 100

34. If two logs are 0.60 and 0.60 –3, we understand that the first log (0.60) has a characteristic of 0, and we then could write the subtraction thus: 0.60 –0, with the
<u>0.60 –3</u>

larger value on top. We need to subtract just the characteristics; change the sign of the –3, and combine with the 0 to get ______ . 3

35. Since the difference in the logs is 3, we know from the antilog of 3 that the numbers having these logs are in the ratio of ______ to 1. 1000

36. If the logs of the numbers are 0.94 –2 and 0.94 –4, the difference in the logs is ______ . 2

37. If the logs we want to subtract differ in *both* the mantissas and the characteristics, we do *two* subtractions, and combine the result. For example, suppose the logs are 0.60 –1 and 0.30 –2. Set up the subtraction: 0.60– 1. Subtract the mantissas to get
<u>0.30 –2</u>

0.30. Subtract the characteristics to get ______ . 1

38. We put together the difference in the characteristics (which was 1) and the difference in the mantissas (which was 0.30) to get a total difference of 1.30. The antilog of 1.30 is the ratio of the two numbers having the original logs. The ratio is ______ to 1. 20

39. Similarly, if the logs are 0.90 –2 and 0.30 –3, the difference in the characteristics is ______ . 1

40. For the same logs, the difference in the mantissas is ______ . 0.60

41. The complete difference in the logs is the combination of the two differences we just found, or ______ . 1.60

42. The entire difference in these logs: 0.60 –1 and 0.20 –3 is ______ . 2.40

43. You may need to subtract two logs like 0.30 –1 and 0.60 –2. You see that both the

mantissas and the characteristics are different. We write the subtraction: 0.30 –1,
0.60 –2

again with the *larger* log on top. The mantissas cannot be easily subtracted as they stand, so we resort to this trick: increase the upper mantissa by 1 and make it 1.30. Now subtract 1 from the upper characteristic, and make it –2. (The net value of the upper log is not changed—it has gained 1 and lost 1.) The upper log is now 1.30 –2. Now the setup reads: 1.30 –2, and the difference in the left-hand pair of values
0.60 –2

is ______ . — 0.70

44. The difference in the changed mantissas is 0.70; since the transformed characteristics are the same, their difference is 0, and thus the entire difference in the logs is just ______. — 0.70

45. The antilog of the difference just found is the ratio of the numbers having the original logs. The antilog of 0.70 is ______ . — 5

46. We follow the same procedure whenever the mantissa of the smaller log is bigger than the mantissa of the larger log. For the two logs 0.70–1 and 0.90–2, we do the same thing: the setup is: 0.70 –1. Increase the top mantissa by 1 to make it
0.90 –2

______ . — 1.70

47. Now increase the top characteristic by –1 to make it ______ . — –2

48. Now write the subtraction: 1.70 –2. The difference is ______ . — 0.80
0.90 –2

49. If the logs are 0.30 and 0.70 –1, the setup is 0.30 . Use the same method as
0.70 –1

before, and find the difference to be ______. — 0.60

50. If the logs are 0.20 –1 and 0.90 –2, the difference in the logs is ______ . — 0.30

51. If the logs are 0.7 –3 and 0.48 –1, write the setup with the *larger* log on *top*: 0.48 –1. Increase the top mantissa by 1, and make the top characteristic also 1
0.78 –3

more in the negative direction. Now you have for the upper log ______ . — 1.48 –2

52. The changed setup reads: 1.48 –2. Now subtract the changed mantissas on the left
0.78 –3

to get ______ . — 0.70

53. Subtract the characteristics as you did before (item 28): change the sign of the bottom characteristic to +, and combine the +3 with the –2 to get______ . — (+) 1

54. The difference in the mantissas was 0.70 and the difference in the characteristics is 1.00. Together they make a total of ______ . — 1.70

55. You found the difference in the two logs to be 1.70 altogether. The antilog of 1.70 is the ratio of the numbers having the original logs. The ratio is______to 1. — 50

56. To subtract these logs: 0.99–3 and 0.45–1. Set up the subtraction, with the larger log on top: ______ . — 0.45 –1
0.99 –3

57. Increase the top mantissa by 1 and its negative characteristic by the same amount, to get the new setup: ____________.

    1.45 −2
    0.99 −3

58. Subtract the changed mantissas to get ____________ . — 0.46
59. Subtract the characteristics to get ____________ . — (+) 1
60. Combine the difference in the mantissas and in the characteristics to get the entire difference ____________ . — 1.46
61. Take the antilog of the 1.46 to get the ratio of the numbers belonging to the logs we started with; it is ____________ to 1. — 29
62. If the logs of two numbers are 0.60 −4 and 0.30 −1, the difference in the logs is ____________ . — 2.70
63. The antilog of 2.70, and thus the ratio of the numbers having those logs, is ____________ to 1. — 500
64. If two numbers have logs that are 0.75 −2 and 0.55, follow the same procedure to get a difference of ____________ . — 1.80
65. You have seen how to find the difference in a pair of logs of decimal fractions if the logs are written in the form mantissa–characteristic. If the logs are given in bar notation, change them to the first form and proceed as before. For example, if the logs are $\overline{1}.4$ and $\overline{2}.2$, change them to 0.4 −1 and 0.2 −2 and proceed with the subtraction, to get ____________ . — 1.2
66. If the logs are $\overline{2}.8$ and $\overline{1}.4$, the difference is ____________ . — 0.6
67. If the logs are $\overline{3}.75$ and $\overline{1}.25$, the difference is ____________ . — 1.50
68. For logs given in minus-ten notation, the subtraction is usually straightforward. For the logs 9.78 −10 and 8.90 −10, the setup is:

    9.78 −10
    8.90 −10

    Since the minus-tens are the same in both logs, they do not affect the difference. We need only subtract 8.90 from 9.78 to get the result, which is ____________ . — 0.88
69. If the logs are 7.60 −10 and 8.30 −10, their difference is ____________ . — 0.70
70. If the logs are 9.50 −10 and 7.90 −10, their difference is ____________ . — 1.60
71. The ratio of the numbers having the preceding logs is ____________ to 1. — 40
72. If the logs are 0.30 and 9.10 −10, set up the subtraction:

    0.30
    9.10 −10

    Follow a procedure like that used before: Increase the upper mantissa by *10* and follow with −10 to keep the value of the log the same. We have:

    10.30 −10
    9.10 −10

    and now subtract as before to get ____________ . — 1.20

SELF-TEST ON THE DIFFERENCE BETWEEN TWO LOGARITHMS

Check your understanding of this section by answering the following questions. The correct answers follow. For each of the following pairs of logarithms, find the difference in the logs and enter in the second column. From the log difference, find the ratio of the numbers having the original logs, and enter in the third column.

	Logarithms	*Difference*	*Ratio of Numbers*
1.	1.5, 1.2	________	________ to 1
2.	2.70, 2.00	________	________ to 1
3.	1.00, 2.60	________	________ to 1
4.	0.7 –1, 0.4 –1	________	________ to 1
5.	0.5 –1, 0.5 –2	________	________ to 1
6.	0.5 –3, 0.5 –1	________	________ to 1
7.	0.9 –2, 0.2 –3	________	________ to 1
8.	0.2 –2, 0.9 –3	________	________ to 1
9.	0.9 –2, 0.6	________	________ to 1
10.	$\bar{1}.7$, $\bar{2}.4$	________	________ to 1
11.	9.7 –10, 8.4 –10	________	________ to 1

ANSWERS TO SELF-TEST ON THE DIFFERENCE BETWEEN TWO LOGARITHMS

In the parentheses after each answer you find the items of this section that relate to each question and its answer.

1. 0.3 (items 1, 4); 2 to 1 (items 2, 3, 6).
2. 0.7 (item 9); 5 to 1 (items 7, 10).
3. 1.6 (item 11); 40 to 1 (items 12, 13).
4. 0.3 (items 18–19); 2 to 1 (item 20).
5. 1 (items 28, 30); 10 to 1 (item 29).
6. 2 (item 32); 100 to 1 (item 33).
7. 1.7 (items 37–38; 50 to 1 (item 38).
8. 0.3 (item 43); 2 to 1 (items 2, 3, 6).
9. 1.7 (items 51–53); 50 to 1 (item 38).
10. 1.3 (item 65); 20 to 1 (item 38).
11. 1.3 (item 68); 20 to 1 (item 30).

SECTION A6
CONVERTING LOGARITHMS OF FRACTIONS

In Section A4 you encountered three different ways of writing the log of a fraction. For example, the logarithm of 0.4 is: 0.60 –1, or $\bar{1}.60$, or 9.60 –10. The logarithm, so written, consists of *two* numbers, a positive mantissa and a negative characteristic. In fact, such a logarithm is an indicated subtraction, not as usually written completed. In some applications it is useful to carry out the subtraction and to find a *single* number equal in value to the *pair* of numbers—one positive and one negative. If you have a pocket calculator that gives the logs of numbers, you will need to perform the reverse calculation.

WHAT YOU WILL LEARN

If you work carefully with the following items you will be able to:

1. Change the logarithm of a fraction into a single negative equivalent value;
2. Change the single negative value of a logarithm into any one of the equivalent forms.

DIRECTIONS. Cover about 2 inches of the right-hand margin of each of the following pages in turn with an opaque sheet of paper. Read carefully the first numbered statement. *Write* the word or phrase that you believe correctly completes the statement. Move the cover sheet down to reveal the correct answer in the margin. Continue in this way until you have finished the section. Work at a rate that is comfortable for you, and no longer than you find pleasant. Begin when you wish.

1. For example, the logarithm of 0.2 is 0.30 –1. If we complete the subtraction, we take 0.30 from 1.00, find 0.70, and prefix the result with a negative sign, and get ________ . —0.70
2. The result (–0.70) is the *single negative* value equivalent to the original 0.30 –1. An analogy is this: if you have 30 cents but owe a dollar, your net "worth" is that you are in the hole by ________ cents. 70
3. Similarly, the log of 0.05 is 0.70 –2. The net value of this pair of numbers would be found by taking the 0.70 from 2.00 and putting a negative sign in front of the result to get ________ . –1.30
4. The log of 0.004 is 0.60 –3. Take the 0.60 from 3.00 and put a minus sign in front of the result to get ________ . –2.40
5. The single negative value equal to the log of 0.07 (which is 0.84 –2) is ________ . –1.16
6. If the log is 0.48 –1, the equivalent single negative value is ________ . –0.52
7. If the number has a log given in the bar notation, we follow the same procedure. In bar notation the log of 0.003 is written as $\bar{3}.48$. To find the single value of the log, we again take the mantissa (0.48) from the characteristic (–3) and prefix the result with a negative sign. We get ________ . –2.52
8. In the same way, if the bar notation is $\bar{4}.75$, the single negative value of the log is ________ . –3.25
9. A log in bar notation is $\bar{2}.55$, equal to the single negative value ________ . –1.45
10. If the log is given in minus-ten notation, we follow the same procedure. If the log is 9.70 –10, it is analogous to a situation in which you have \$9.70, but owe \$10—you are behind by ________ cents. 30
11. Thus to arrive at a single number from 9.70 –10, we subtract 9.70 from 10.00 and put a minus sign before the result to get ________ , –0.30
12. If the log is 8.66 –10, subtract 8.66 from 10.00, put a negative sign in front of the result, and get ________ . –1.34
13. If the log is 7.33 –10, the equivalent single value is ________ . –2.67
14. The log 6.782 –10 equals the single value ________ . –3.218

15. Each of the equivalent logs: 0.375−3, $\bar{3}$.375, and 7.375−10 is equal to the single negative value ________. −2.625

16. Many desk and pocket calculators give the log of a fraction directly as a single negative number. It is sometimes useful to be able to change this value into one of the other ways of writing the log of a fraction. Suppose that the log is given as −0.60. We cannot immediately look up this number in the table because the table gives only *positive* mantissas. Thus we need to change −0.60 into a form with a positive mantissa (and a negative characteristic). The characteristic will be the whole negative number just larger than the given value. In this case, −1.00 is that number. The mantissa will be the positive value found by taking the difference (ignoring the minus signs) between the given value 0.60 and the characteristic 1.00. This difference is ________. 0.40

17. Thus the mantissa is 0.40 and the characteristic is −1, and we would write the log as 0.40 −1, or $\bar{1}$.40, or 9.40 −10. By the same process, we transform −0.75 into a log with the characteristic ________. −1

18. Continuing, we take 0.75 from 1.00 to get the mantissa ________. 0.25

19. Thus −0.75 is equal to 0.25−1, or $\bar{1}$.25, or 9.25−10. By the same operations, −0.12 is the same as ________ (mantissa − characteristic). 0.88 −1

20. If the negative log is −1.7, the whole negative number just larger than the given value is ________. −2

21. For the given log −1.7 the characteristic is −2, and the mantissa, as before, is found by subtracting 1.7 from 2.0, to get ________. 0.3

22. Transformed, −1.7 is the same as 0.3 −2. Similarly transformed, −1.43 is the same as ________. 0.57 −2

23. In bar notation, −2.4 is the same as ________. $\bar{3}$.6

24. If the log is given as −1.861, in bar notation the log is ________. $\bar{2}$.139

25. If the log is −4.656, in bar notation the log is ________. $\bar{5}$.344

SELF-TEST ON CONVERTING LOGARITHMS OF FRACTIONS

Check your understanding of this section by answering the following questions. The correct answers follow.

1. The single negative logarithm equivalent in value to the following logarithms is:
   (a) 0.3 −1 ________.
   (b) 0.6 −3 ________.
   (c) 0.68 −2 ________.
   (d) $\bar{3}$.48 ________.
   (e) $\bar{2}$.15 ________.
   (f) 9.70 −10 ________.
   (g) 8.45 −10 ________.

2. Transform each of the following negative logarithms into the form: mantissa – characteristic.
   (a) –0.60 ____________ .
   (b) –0.75 ____________ .
   (c) –1.70 ____________ .
   (d) –2.134 ____________ .
3. Write each of the answers to question 2 in bar notation.
   (a) ____________ .
   (b) ____________ .
   (c) ____________ .
   (d) ____________ .

ANSWERS TO SELF-TEST CONVERTING LOGARITHMS OF FRACTIONS

In the parentheses after each answer you find the items of this section that relate to each question and its answer.

1. (a) –0.7 (items 1, 2).
   (b) –2.4 (item 4).
   (c) –1.32 (items 1–6).
   (d) –2.52 (item 7).
   (e) –1.85 (items 7–9).
   (f) –0.30 (items 10–11).
   (g) –1.55 (item 12).

2. (a) 0.40 –1 (item 16).
   (b) 0.25 –1 (items 17–19).
   (c) 0.3 –2 (items 20–21).
   (d) 0.866 –3 (items 21–22).

3. (a) $\bar{1}.40$ (items 23–25).
   (b) $\bar{1}.25$ (items 23–25).
   (c) $\bar{2}.3$ (items 23–25)
   (d) $\bar{3}.866$ (items 23–25).

APPENDIX B

Low-cost Sensitometry

If conventional equipment is used a sensitometric test laboratory may cost several thousands of dollars. It is possible, however, to obtain useful information about the photographic response of sensitive materials without expensive apparatus. Only two requirements are necessary:

1. A method of supplying the sample material with known exposures, or at least exposures of known relationship;
2. A method of measuring the resulting image.

In this section we describe techniques for providing these two requirements without expensive equipment.

HOW TO EXPOSE THE SAMPLE

Construct a Sensitometer

A. Use as a light source a tungsten lamp at a suitable distance from the exposure plane. Or, use a contact or projection printer. All that is needed is a reasonably stable source at a manageable level.

   Measure the light at the exposure plane with an illuminance meter (see Appendix C). Check the uniformity of light over the exposure plane. If necessary, reduce the illuminance (to permit the use of a reasonable exposure time) by using neutral filters or apertures.

of a reasonable exposure time) by using neutral filters or apertures.

B. Use a step tablet as the test object. Uncalibrated step tablets with a range in logs of about 3, with 20 increments of about 0.15, can be purchased for a few dollars. Place the step tablet at the exposure plane or in the negative carrier of a projection printer, as desired.

C. Regulate the time of exposure. Use a shutter or (if the illuminance is low enough) merely turn on and off the lamp. If electronic flash is used as a source, the exposure time (if not accurately known) is nearly the same for each exposure.

NOTE. This method is preferable to those that follow because it can give "absolute" values of exposure. The light source, however, may be remote from that used in practice. Furthermore, if long times of exposure are used, reciprocity effects may be important.

Use a Camera and a Reflection Gray Scale

In this case, only the *relative* log exposure values are known from the preprinted numbers on the gray scale. Be sure that the gray scale is evenly lighted and that no glare reflections result. See that the image of the gray scale lies near the center of the frame, to reduce the error associated with light falloff in the camera. See that the size of the patches in the image is large enough to permit measurements of the patches.

NOTES. This method is close to reality, because it is used with the photographic system of interest. Since the actual exposures on the film are not known, only relative data can be obtained. This method, however, can be used to find out whether or not an assumed exposure index is correct, or to determine filter factors and check the relationship between shutter settings and *f*-numbers, and so on.

Obtain Relative Exposure Values from Subject Meter Readings

This method is the nearest to the real world. Small-spot meters, properly used, can give very reliable data if they are in calibration. The subject areas to be measured must be neutral—that is, gray—unless the meter is designed so as to duplicate the standard human visual response.

HOW TO ESTIMATE NEGATIVE DENSITIES

Construct a Densitometer

You will need a photoelectric light meter of suitable sensitivity and range. There are many models that have a range of over 6 in logs, with scale intervals equivalent to 0.3, with which density values can be measured with a precision of a few hundredths. In addition, you will need a stable light source such as a small tungsten lamp and an aperture having a diameter of a few millimeters. A sketch of the assembly follows:

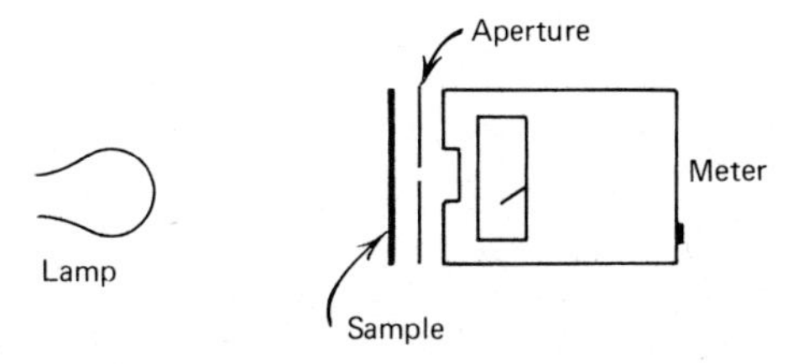

With nothing over the aperture the reading may be 15; this is the zero position. With a sample in place the reading may be 7. The difference from the zero setting is 8 scale divisions; since each interval is 0.3, the density is 8 × 0.3 or 2.4.

Use a Set of Known Densities for Visual Comparison

The known densities may be a step tablet like that used for exposing a sample. Such a tablet contains a series of densities from 0.05 to 3.05, in increments of 0.15.

It is necessary to place the known density areas next to the ones to be estimated. Therefore see that the image (e.g., of a reflection gray scale) is at the margin of the negative. Trim the negative so that the image to be measured is at the extreme edge. Place a step tablet (also trimmed) alongside the image. Cut an aperture in a card so that you can see only the areas being compared. Place the assembly over a diffuse light source, such as an illuminated ground glass. Slide the step tablet along until you obtain a visual match. A sketch of the arrangement is as follows:

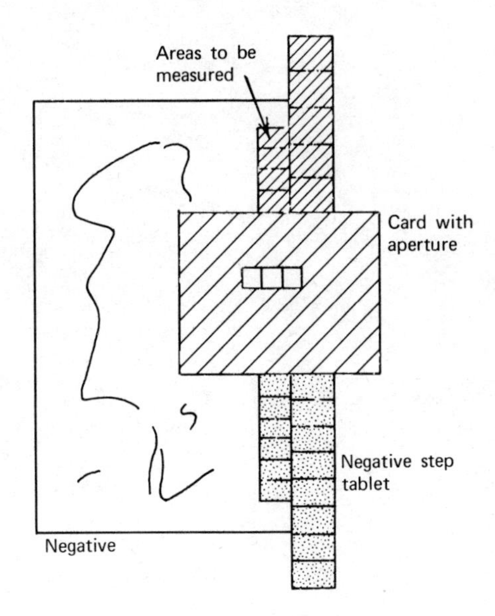

Interpolation between steps can give values correct to within a few hundredths. A disadvantage of the method is that the image to be measured must be at the margin of the negative; thus there is a light falloff problem in the exposure of the edge of the negative.

Print densities may be measured with a similar arrangement, using as a reference a reflection gray scale that has marked density values.

APPENDIX C

Measurement of Light Sources

Photographers measure light with meters and expect that the meter readings will aid them in setting their cameras so as to obtain well-exposed negatives. This programed text deals with the fundamentals of *photometry,* that is, the measurement of light.

By the term "light" we mean that kind of radiant energy that enables our eyes to function. This definition excludes radiation similar to light but not visually effective; it specifically excludes infrared and ultraviolet radiation. By this definition, basic measurements of light necessarily involve the eye. Photoelectric meters can be substituted for the eye only, in general, if they "see" radiation as does the eye.

Much of the following is concerned with vocabulary—the terms we use for the measurement of light. This vocabulary has been developed historically. Just as we measure the performance of an engine in horsepower, so we use the term "candlepower" as a measure of a light source, even though the candle is long gone as a basis of light measurement. Partly for historical reasons, the terminology used in light measurement is troublesome. The following text is intended to help you make sense out of an often confusing situation.

There are two different situations in which measurements of the quantity of light are made:

1. The output of a source, such as a tungsten lamp. Since a reflecting surface like the face of a model acts as a source of light to the camera,

similar measurements are used to measure the light reflected *from* a subject.

2. The light falling *on* a surface, which may be a subject of a photograph, or a film in a camera or a sheet of photographic paper on the easel of a projection printer.

Different methods of measurement and different terms are used in these two different situations. The two are, however, connected; if you know the output of a lamp and its distance from a surface, you can often know the strength of the light falling on the surface.

SECTION C1 INTENSITY

If you work carefully with the material in the following section you will learn:

1. What is meant by the "candlepower" of a lamp;
2. That the intensity of a lamp changes, depending on the direction of view;
3. That the use of a reflector with a lamp changes its effective intensity in a desired direction;
4. How information about the intensity of a lamp is described by a special type of graph.

DIRECTIONS. Cover about 2 inches of the right-hand margin of each of the following pages in turn with an opaque sheet of paper. Read carefully the first numbered statement. *Write* the word or phrase that you believe correctly completes the statement. Move the cover sheet down to reveal the correct answer in the margin. Continue in this way until you have finished the section. Work at a rate that is comfortable for you, and no longer than you find pleasant. Begin when you wish.

We define the output of a lamp by comparing it with a standard source of light. The standard used to be a candle. Because real candles are impossibly variable, this standard has been replaced by a more stable source of light. It is now a cavity in a block of graphite heated to a temperature high enough to glow. To emphasize this replacement, the term "candela" is slowly replacing the older word "candle."

1. When we speak of the candlepower of a lamp, we specify its *intensity*. Imagine a candle flame shrunk into a tiny ball. It would produce light equally in all direc-

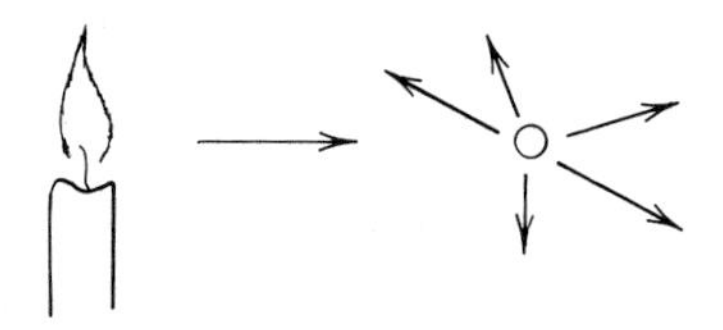

tions. Five candle flames shrunk into a similar sphere would be the equivalent of a ___________ candlepower source. 5

2. A 100-watt tungsten household lamp has an intensity of about 1.25 candelas for every watt. The intensity (candlepower) of such a lamp is about ________ candelas. — 125

3. A 50-candlepower lamp has twice the intensity of a ________ -candlepower lamp. — 25

4. For real sources, the intensity varies with the direction. Merely by putting a mirror behind a lamp, we increase the intensity in some direction *A*, so that the original 1-candlepower lamp now has an effective intensity of ________ candelas in direction *A*. — 2

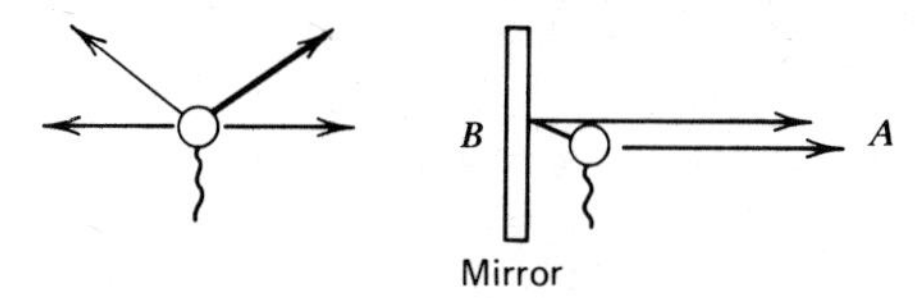

5. In direction *B* the intensity of the lamp-mirror combination would be ________ . — 0

6. The effective change in intensity of a lamp is the purpose of using a reflector with a flashbulb or a studio lamp; the effect is to increase the ________ of the lamp in a desired direction. — intensity or candlepower

7. Even without a reflector a real lamp will have different intensities in different directions. A clear-bulb tungsten lamp may have a filament in the shape of a small cylinder. The intensity of the lamp would be large in direction *A* because the

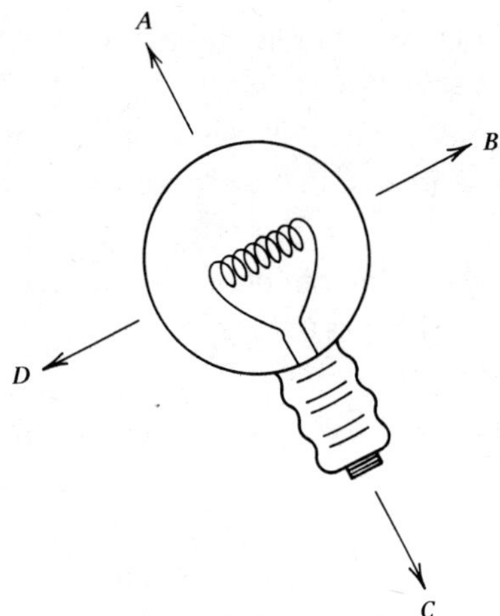

filament has a large light-producing surface area in that direction. On the other hand, in direction *B* the intensity would be ________ . — small

8. Because the base of the lamp gets in the way, the intensity would be practically zero in direction ________ . — C

9. The intensity of the lamp would be about the same in directions ________ and ________ . — B, D

10. The preceding items illustrate that the intensity of a real lamp changes with direction. To specify the intensity variation of a lamp, a graph may be plotted on graph paper like that below. The paper has labeled angles, just like a circular

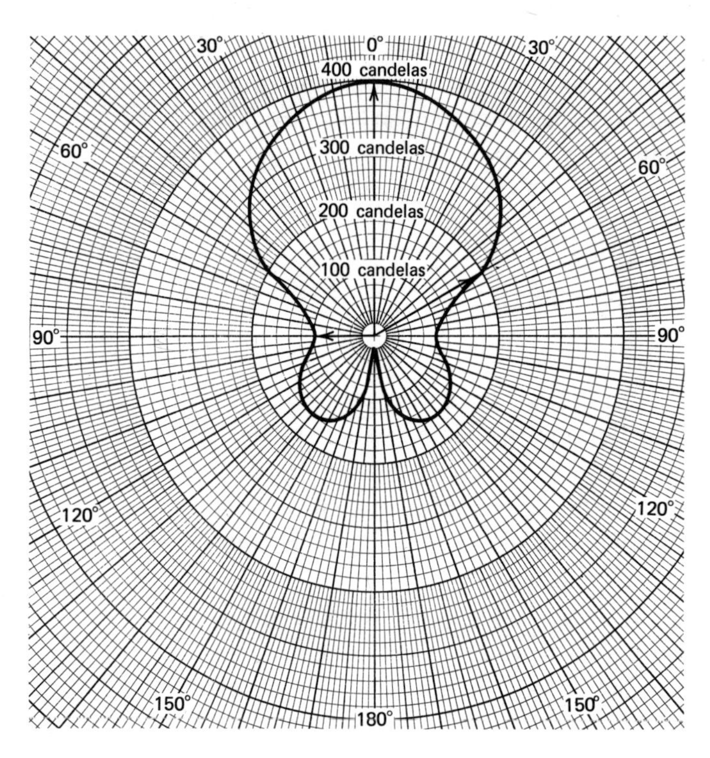

protractor, on which are superimposed circles of varying radius. The circles are identified with numbers, in this case different candlepower values.

The graph above (for the lamp sketched in item 7) shows the intensity in any desired direction, as shown by the lengths of the arrows. In the 0° direction, the arrow length is __________ (candelas). — 400

11. Thus the intensity of the lamp is 400 candelas in this direction. At 90° (i.e., directions *B* and *D* in item 7) the candlepower of the lamp is only__________. — 100 cd.

12. At 60° the intensity of the lamp is __________ . — 200 cd.

13. At 180°, because of obstruction by the base of the lamp, the intensity is practically __________ . — 0 cd.

14. Because this lamp has a great variation in intensity with direction, in use (as in a slide projector) it would be important to have the filament properly oriented in the lamp housing. It is for this reason that most lamps used for such a purpose are provided with pin or similar bases so that they can be correctly placed in the socket.

An ideal (''perfectly uniform) source would have the same intensity in all directions. The graph of such a source on this type of graph paper would be a __________ . — circle

15. A circular graph (with center of the circle at the center of the graph paper) would describe a uniform spherical source. Every real lamp gives at least *some* variation

in intensity in different directions, and thus the graph for a real source is never a perfect circle. The *maximum* intensity of the lamp is called the *beam* candlepower. For the lamp described by the plot in item 10, the beam candlepower is ________ .

400 cd.

16. When a single value of intensity is given for a lamp, the value is almost always the maximum, that is, the ________ candlepower.

beam

17. A manufacturer advertises an 80,000 candlepower flashlight. The number is almost certainly the ________ candlepower.

beam

18. Such a large intensity can be obtained by putting the lamp bulb in front of a curved reflector, which causes most of the light to go in a narrow beam and thus increases ________ of the lamp in one direction.

intensity or candlepower

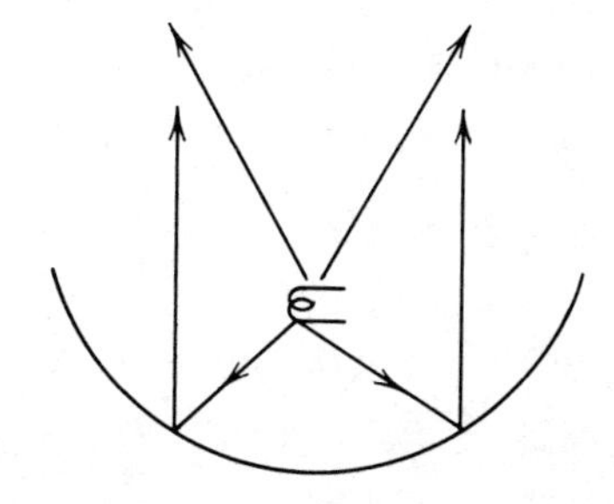

19. The intensity of the lamp in most other directions would be ________ candelas.

0

20. The plot of intensity versus direction for this lamp-reflector combination would be as shown in the following diagram. Here only half the graph paper is used because

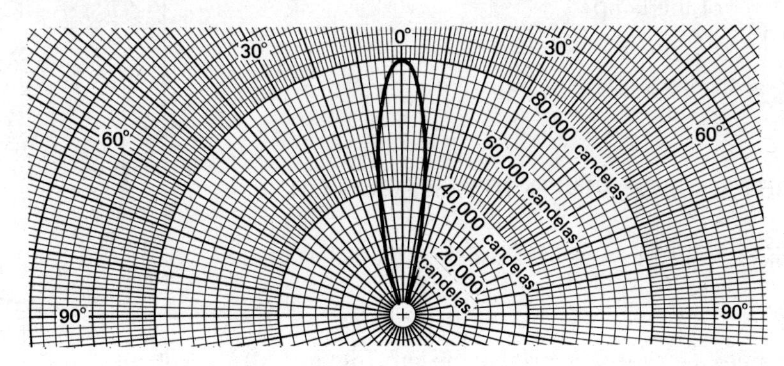

the reflector permits no light to go downward. From the beam candlepower value of 80,000 candelas, the intensity falls to half that value at an angle of ________ °.

10

21. Beyond an angle of about 20° the lamp has an intensity of practically ________ .

0 candela

22. For use *at* the camera, a photographic lamp should cover the angular field of the camera lens with near uniformity. If we assume a lens with a total field angle of 50°, a suitable lamp would be described by the graph that follows:

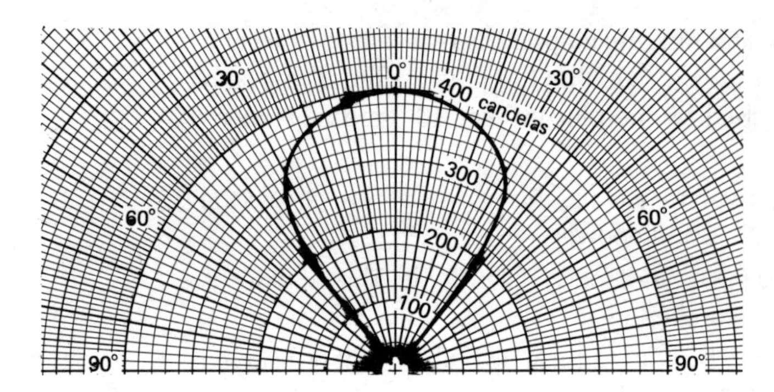

The beam candlepower of this lamp is ___________ .

400 candelas

23. At an angle of 25° (half the total field angle of 50°) the intensity falls to about ___________ . Thus the intensity variation over the whole field of the lens (an angle of 50°) would be from 400 candelas to 360 candelas, a change of only 10%—less than would usually be noticeable.

SELF-TEST ON LIGHT SOURCES—INTENSITY (SECTION C1)

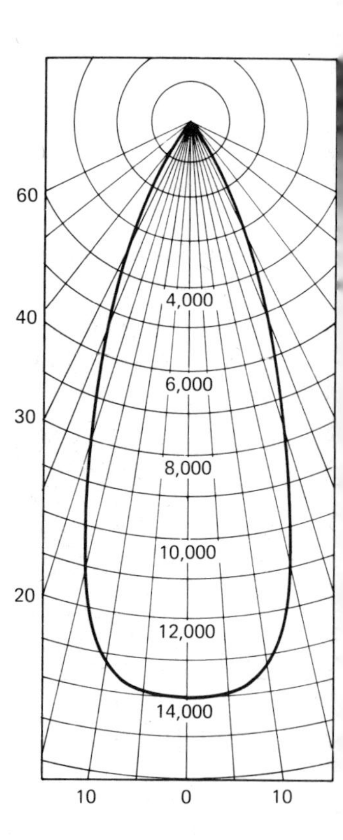

Check your understanding of this section by answering the following questions. The correct answers follow.

1. What characteristic of a lamp is described by the term "25 candlepower"?
2. Why is the single measure of this characteristic insufficient?
3. Refer to the graph of the lamp at the right:
   (a) What is the intensity of the lamp head on, in other words, at an angle of 0°?
   (b) What is the beam candlepower of the lamp?
   (c) What is the intensity of the lamp at an angle of 25°?
   (d) By what percentage does the lamp intensity change from the angle of 0° to 25°?

ANSWERS TO SELF-TEST ON LIGHT SOURCES—INTENSITY (SECTION C1)

In the parentheses after each answer you find the items in this section that relate to the question and its answer.

1. The term "25 candlepower" identifies a value of *intensity* of the lamp (items 1–3).
2. The single number is inadequate because the intensity of a lamp is different in different directions (items 4–14).
3. (a) The intensity at 0° is nearly 14,000 candelas (item 10).
   (b) The beam candlepower is the same number as in (a), since beam candlepower is the maximum intensity (items 15–22).
   (c) The intensity of the lamp at 25° is only 4000 candelas (items 10–13).
   (d) At 25° the intensity is only about 27.5% of what is at 0° (item 23).

SECTION C2
MEASUREMENT OF LIGHT SOURCES—LUMINANCE

If you work carefully with the following sections, you will learn:

1. That the observed brightness of a lamp is related to the *luminance* of the lamp;
2. That luminance is a measure of the intensity of a unit area of a source;
3. That luminance is measured with meters ordinarily called "reflectance" or "reflected-light" meters;
4. That when the subject is considered to be the source of light for the camera, subject luminances are directly connected to the exposures on the film.

DIRECTIONS. Cover about 2 inches of the right-hand margin of each of the following pages in turn with an opaque sheet of paper. Read carefully the first numbered statement. *Write* the word or phrase that you believe correctly completes the statement. Move the cover sheet down to reveal the correct answer in the margin. Continue in this way until you have completed the section.

1. A 100-watt clear tungsten bulb and a 100-watt frosted bulb have about the same average intensity measured in ________ . — candelas, or candlepower
2. The two lamps, however, look much different. The clear bulb appears much *brighter* than the frosted bulb. The reason for the difference in appearance is that for the clear lamp the source is the tiny filament, whereas for the frosted lamp the effective source is the much larger glass bulb itself. Because the frosted lamp has a larger surface area, its brightness is ________ . — smaller, less (etc.)
3. By the term "brightness" we mean the appearance of the source as related to the *visual* impression of glare, brilliance, or harshness. The sun is a source of very great ________ . — brightness
4. For a given intensity (candlepower) the brightness is *less* as the area of the source is made ________ . — more, or greater
5. The visual brightness is related to the intensity (candlepower) of a *unit area* of a source of light. Suppose that the area of the filament of the clear bulb is 1 cm^2, and its intensity is 125 candelas. For the frosted bulb of the same intensity the light is coming from a much larger area, say 125 cm^2. *Each* square centimeter of the frosted lamp would have an intensity of only ________ . — 1 candela
6. The difference between the two lamps is sketched as follows:

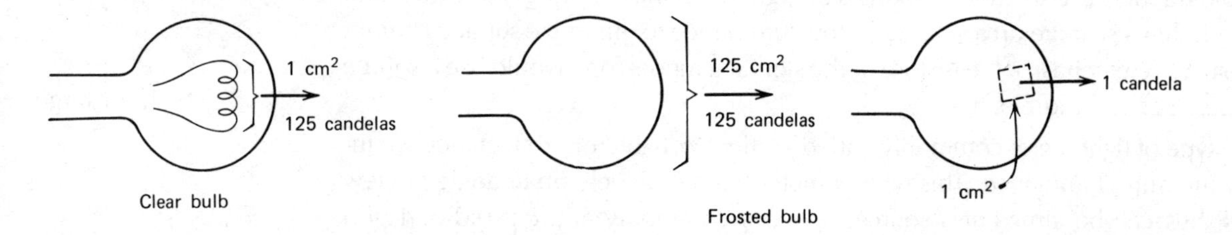

Brightness—the visual appearance—is closely related to the intensity per *unit area* of the source. Unit area—1 cm^2—of the *clear* lamp has an intensity of ________ candelas. — 125

7. One cm^2 of the *frosted* lamp, however, has an intensity of only ________ . — 1 candela
8. The intensity per unit area is called the *luminance*. The luminance of the clear bulb is 125 candelas per square centimeter, usually written 125 cd/cm^2. The luminance of the frosted bulb is ________ cd/cm^2. — 1
9. If a 500-candlepower lamp has an effective area of 10 cm^2, the luminance is ________ cd/cm^2. — 50
10. You will see luminance expressed as the number of candelas for other measures of area, such as candelas per square meter (cd/m^2) or candelas per square foot (cd/ft^2). Since there are (100 × 100) square centimeters in a square meter, a luminance of 1 candela per square centimeter is equal to ________ cd/m^2. — 10,000
11. For a clear 500-watt tungsten lamp compared with the same lamp in a frosted globe, the average ________ is the same. — intensity
12. For the lamps in item 11, the clear bulb would have the greater ________. — luminance
13. The distinction between intensity and luminance is that intensity applies to the *whole* lamp, whereas luminance applies only to unit ________ of the lamp. — area
14. A fluorescent lamp has a total intensity of 1000 candelas in a given direction. Its area as seen from the same direction is 250 cm^2. Its luminance is ________ ____ .
    number unit — 4 cd/cm^2
15. A tungsten lamp has the same intensity as the fluorescent lamp in item 13, but its area is only 20 cm^2. Its luminance is ________ ____ .
    number unit — 50 cd/cm^2
16. Tungsten lamps are used in table transparency viewers, and are placed behind a sheet of groundglass. The effect of the groundglass is to make the effective area greater and thus to make the luminance ________ . — less, or smaller
17. A low-luminance source used as a photographic lamp produces soft, diffuse light on the subject. On the other hand, hard (contrasty) lighting is produced by sources of small area, and thus of ________ luminance. — large (etc.)
18. When a photographer uses "bounce" lighting, he aims his flashlamp at a ceiling (e.g.,); in this case the ceiling becomes the effective light source for the subject. Since the ceiling has a very large area, its ________ will be small. — luminance
19. In bounce lighting, with an effective source that has a very low luminance and a very large area, the lighting on the subject will be very ________.
    soft, hard — soft
20. Photographers use small-area sources of high luminance to bring out texture and detail; they use large-area sources of low luminance to minimize surface characteristics. To emphasize the texture of the skin of a model, one would use a source of ________ luminance. — high, or large
21. The type of light meter commonly called "reflected-light" or "reflectance" actually measures *luminance*. This type of meter has a relatively *small* angle of view and thus can be aimed at a source, or more commonly at a desired part of a subject. The subject is thought of as the source of light for the camera.

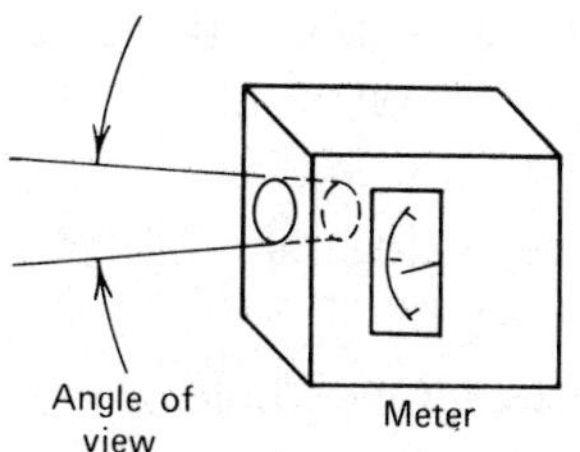

When from the camera position the photographer points such a meter at his subject, he is measuring its __________ .

luminance

22. In the United States the unit of luminance is usually __________ per square foot.

candles or candelas

23. European meters read in __________ per square meter.

candles or candelas

24. The actual numbers are usually arbitrary values with intervals related to logarithms, for reasons that have been developed in Chapter 2, items 69–92. The numbers on the scale are for this type of meter dependent upon the __________ of the subject.

luminance

25. Different parts of the subject that have different visual brightnesses give different luminances when measured with a meter of restricted angle of view. Suppose that one part of a subject (perhaps the forehead of a model) has a luminance 10 times that of another part (perhaps the dark hair of the model). Each unit area of the forehead would send toward the camera __________ as much light as the hair.

10 times

26. The situation is sketched in the following diagram.

1 cd/cm^2

10 cd/cm^2

Camera

If there is little scattered light in the camera optical system—little *flare*—the luminance *ratio* will be preserved in the image as seen on the groundglass of a view camera or in the viewer of a reflex camera (e.g.). What is more important is that nearly the same ratio will be found for the exposures for the forehead and hair *on the film* in the camera. For the situation above, the exposure ratio would be nearly __________ to 1.

10

27. Thus there is an expectable near-equality between the ratio of exposures on the film and the ratio of the measured __________ of the subject.

luminances

28. If we take the logarithm of this ratio, we expect that the result is nearly the logarithm of the exposure interval on the log *H* axis of the *D*–log *H* curve. Here the log interval is __________ .

1

29. We can reasonably expect that if the luminance ratio of two tones of the subject is 10 to 1 the interval between the log *H* values produced by these tones is __________ .

1

30. Typical outdoor subjects have a highlight luminance that is about 100 times the darkest shadow luminance. For such a subject, the exposure on the film will be for the highlight about __________ times that for the darkest shadow. — 100

31. The log H interval for a typical outdoor subject will be the logarithm of 100, or __________ . — 2

32. A very flat subject may have a luminance ratio of only 20:1. One would expect the ratio of exposures on the film to be about __________ to 1, and the log H interval on the horizontal axis of the D–log H curve to be about __________ . — 20, 1.3

33. Predictable relationships exist between the exposures on the film in the camera and the measured __________ values of the subject. — luminance

SELF-TEST ON MEASUREMENT OF LIGHT SOURCE—LUMINANCE (SECTION C2)

Check your understanding of this section by answering the following questions. The correct answers follow.

1. Lamp *A* has an intensity of 100 candelas and an area of 4 cm^2. What is its luminance?
2. Lamp *B* has an intensity of 100 candelas and an area of 10 cm^2. What is its luminance?
3. Compare the visual brightnesses of lamps *A* and *B*.
4. Lamp *C* has a luminance of 10 cd/cm^2 and an area of 20 cm^2. Compare the intensity of lamps *B* and *C*.
5. What characteristic of a source of light is actually measured by "reflected-light" meters?
6. What is the main feature of construction of the meters mentioned in 5?
7. In a particular scene the greatest luminance is 200 cd/m^2 and the smallest is 2 cd/m^2. What is the luminance ratio?
8. For the same data as in 7, and assuming insignificant flare:
   (a) What is the ratio of exposures in the camera;
   (b) What is the log H interval in the camera.

ANSWERS TO SELF-TEST ON MEASUREMENT OF LIGHT SOURCE—LUMINANCE (SECTION C2)

In the parentheses after each answer you find the items in this section that relate to the question and its answer.

1. 25 cd/cm^2 (items 5–9).
2. 10 cd/cm^2 (items 5–9).
3. Lamp *A* is brighter, in fact about 2½ times as bright as *B* (items 2–5).
4. Lamp *C* has an intensity of 10×20, or 200 candelas; it is thus twice as intense as *B* (item 13).

5. "Reflected light" meters measure luminance (item 21).
6. The most important feature of luminance meters is that they have a restricted (narrow) angle of view (items 21, 25).
7. The luminance ratio is 200:2, or 100:1 (items 25–26).
8. (a) The same as the luminance ratio (i.e., 100:1) (items 26, 27, 30);
   (b) the logarithm of 100, or 2.0 (items 31, 32).

SECTION C3
MEASUREMENT OF LIGHT SOURCES—FLUX

Manufacturer's data about flashlamps and electronic flash usually include a number that estimates the total light output from the lamp without regard to direction. An example of such a measure appears in the following plot for a typical flashbulb.

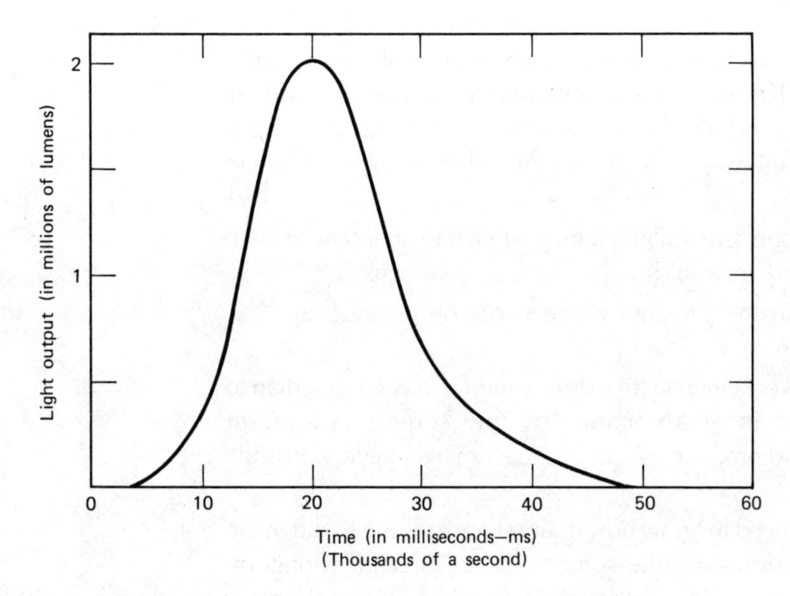

If you work carefully with the following section, you will learn:

1. The meaning of the values—lumens—shown on the vertical axis of such a graph;
2. The relationship between this measure of light sources and intensity.

DIRECTIONS. Cover about 2 inches of the right-hand margin of each of the following pages in turn with an opaque sheet of paper. Read carefully the first numbered statement. *Write* the word or phrase that you believe correctly completes the statement. Move the cover sheet down to reveal the correct answer in the margin. Continue in this way until you have completed the section.

1. Consider a uniform source of 1 candlepower. Its ____________ is the same in all directions. — intensity
2. We intend now to sum into a single value *all* the light produced by the lamp, *regardless* of direction. We first define a unit solid angle, as in the sketch below.

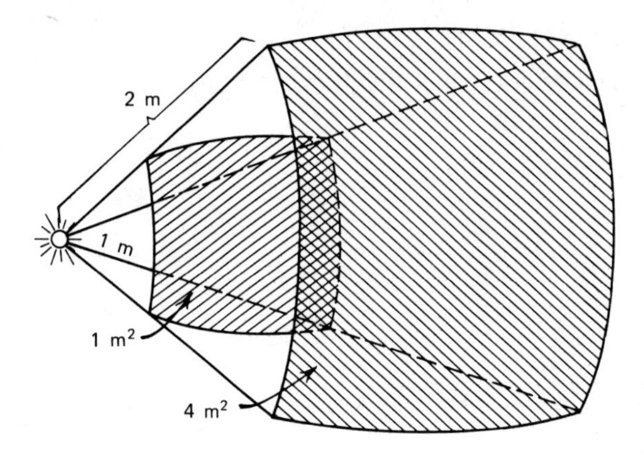

The angle is contained within a pyramid or cone having a base of one square meter area (1 m²) at a distance of 1 meter from the source. The pyramid has an area of 4 m² at a distance of 2 m from the source.

Now think of light as flowing through an aperture equal to the base of the pyramid in the sketch above. If the lamp has an intensity of 1 candela toward the aperture, the light flow—called the "flux"—is 1 *lumen* within this unit solid angle. As the intensity of the source increases, the flux found within a given solid angle also ____________ . — increases

3. The unit solid angle is called the "steradian." If the source had an intensity of 10 candela, the flux in one steradian would be ____________ lumens. — 10
4. Therefore to find the flux within one steradian, we need to know the ____________ of the source in that direction. — intensity
5. We can use the rule that for every candela there is one lumen in each steradian to sum up the light produced in all directions by any source. A *uniform* 10-candlepower source would produce ____________ lumens in every possible steradian. — 10
6. To sum up all of the light, we need to know how many steradians are found in the whole sphere completely surrounding the source. For a 50-candela uniform source, there will be ________ ____ in each steradian. — 50 lumens
   number unit
7. The sketch on page 205 shows that if the radius of the sphere is r, the area of the base of the pyramid defining one steradian is *r squared* (r^2).

   Since the area of the whole sphere is $4\pi r^2$, we can fill the solid sphere with just 4π steradians. Since π is about 3.14, there are 4π, that is, ____________ steradians in the whole sphere. — 12.56
8. To simplify the calculations that follow, call the number of steradians in a whole sphere 12.5. In each of these solid angles, from a 1-candlepower source, there would be ____________ lumen of flux. — 1
9. Therefore from a uniform source of 1 candlepower there will be in the whole sphere a total of about ____________ lumens. — 12.5
10. We now have a conversion factor for a uniform source; to find the number of lumens, multiply the candlepower by 12.5. A 50-watt tungsten lamp has an average intensity of about 60 candelas. Assuming that it is uniform, it produces a total of about ____________ lumens. — 750

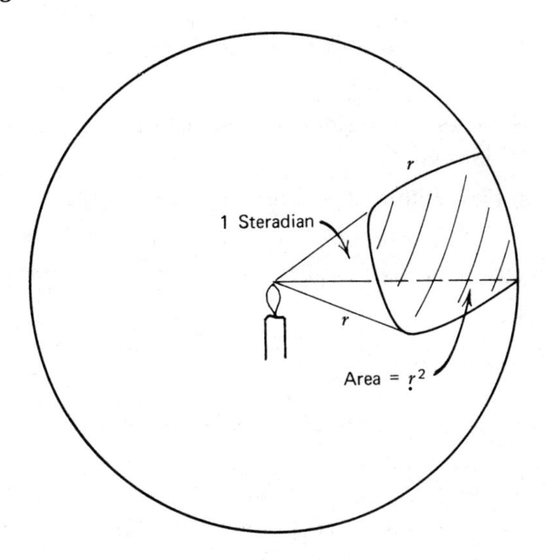

11. From a uniform source of intensity 10 candelas, a total of _______ ___ (number unit) of flux is produced. — 125 lumens

12. Conversely, if you know the total number of lumens from a *uniform* source you can find the candlepower by dividing the number of lumens by the conversion factor 12.5. Thus if such a source produces a total of 25 lumens, it is a __________ candlepower source. — 2

13. Assume the flashbulb having the curve shown in the introduction to this section to be a uniform source. At peak, it produces 3,750,000 lumens. At peak this is a __________ candlepower source. — 300,000

14. No real source is uniform, since it always has an intensity that varies with direction. To find the lumens of flux for such a source, we measure its intensity in each of the 4π solid angles of the sphere, and add up the lumens of flux within each of the angles. The resulting value of flux disregards the __________ in which the light is produced. — direction

15. Consider as an example the light source with a candlepower graph (as in Section C1, items 10–13) shown as follows.

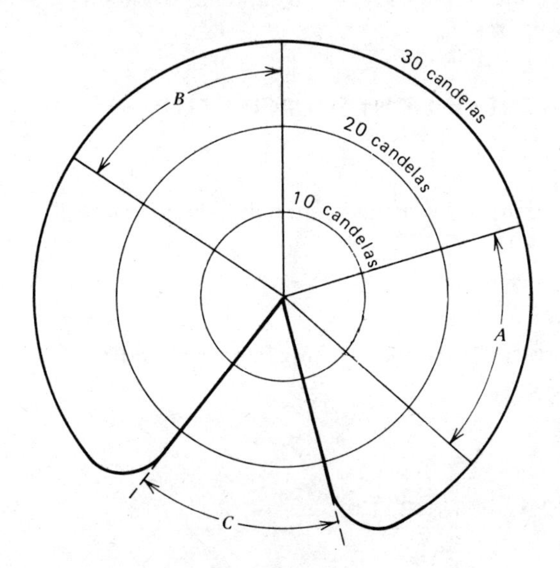

This is a somewhat idealized case; the lamp has in most directions a uniform intensity of ___________ candelas — 30

16. Within the *unit solid angle* indicated by the arc marked *A*, there would be from the 30-candela source a flux of ___________ lumens. — 30
17. There would be another 30 lumens of flux within the solid angle marked ___________ . — B
18. There would similarly be 30 lumens of flux in all the possible 12.5 unit solid angles except for the one marked ___________ . — C
19. There is no flux within the angle marked *C* because the lamp has an intensity of ___________ in direction *C*. — 0
20. Therefore of the 12.5 total number of solid angles, there would be a flux of 30 lumens in 11.5, for a total flux of _______ ___ .
    number unit — 345 lumens

A similar but more complicated process of summing up the number of lumens in the whole sphere would be used for a lamp giving more variation in intensity with direction.

SELF-TEST ON MEASUREMENT OF LIGHT SOURCES—FLUX (SECTION C3)

Check your understanding of this section by answering the following questions. The correct answers follow.

1. In what unit is the flux from a lamp measured?
2. From a lamp of uniform intensity 100 candelas, how many units of flux are found within each unit solid angle?
3. Since there are 12.5 unit solid angles in an entire sphere, what is the total flux from the lamp mentioned in 2?
4. What is the total flux from a uniform lamp of intensity 500 candelas?
5. A flashbulb at its peak produces 1,500,000 lumens of flux. What is its intensity at its peak?
6. A lamp is mounted in a reflector. It has a uniform intensity of 200 candelas over half a sphere. What is its total flux?

ANSWERS TO SELF-TEST ON MEASUREMENT OF LIGHT SOURCES—FLUX (SECTION C3)

In the parentheses after each answer you find the items of this section that relate to the question and its answer.

1. Flux is measured in lumens (items 2–4).
2. A 100-candela lamp, if it is uniform, produces 100 lumens of flux in each steradian (unit solid angle) (items 3–6).
3. 12,500 (i.e., 100×12.5) (items 6–11).
4. 6250 lumens (items 6–11).
5. 120,000 candelas (i.e., $1{,}500{,}000 \div 12.5$) (items 12, 13).

6. In half a sphere there are 12.5/2 steradians (i.e., 6.25).
   For each steradian there will be 200 lumens from a 200-candela source. Thus the total flux is 6.25 × 200, or 1250 lumens (items 14–20).

SECTION C4
MEASUREMENT OF LIGHT SOURCES—QUANTITY OF LIGHT

If you work carefully with the following section you will learn how we measure the total *quantity* of light produced by a source.

DIRECTIONS. Cover about 2 inches of the right-hand margin of each of the following pages in turn with an opaque sheet of paper. Read carefully the first numbered statement. *Write* the word or phrase that you believe correctly completes the statement. Move the cover sheet down to reveal the correct answer in the margin. Continue in this way until you have completed the section.

It is well to think of flux, as defined in the preceding section, as closely similar to the flow of water as from a sprinkler head. Flow is an ongoing process, and thus has the characteristic of a *rate*. Gallons-per-minute is for water flow like *lumens* for light flux.

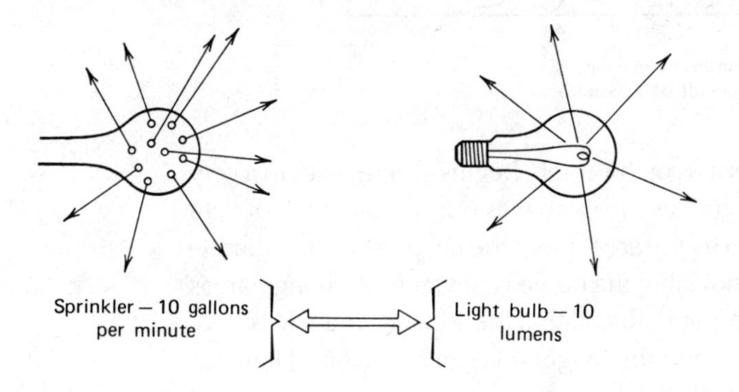

1. The *amount* of water (in gallons) delivered by a sprinkler would depend on the *rate* of flow and the *time* during which the flow goes on. Similarly, the *amount* of light produced by a lamp depends on the flux (in lumens) and the __________ during which the lamp operates. — time
2. For a lamp, the time is usually measured in seconds. The measure of the *quantity* of light from a lamp is *lumen-seconds,* and is found from the *product* of the flux and the time. Thus if a lamp delivers 100 lumens of flux and operates for 5 seconds, the quantity (amount) of light produced during this time is __________ lumen-seconds. — 500
3. An enlarger lamp may produce light at a rate of 3000 lumens. If it is turned on for 10 seconds, the lamp produces a total quantity of ________ ____ .
   number unit — 30,000 lumen-seconds
4. A small laser produces a total flux of 0.1 lumen. To obtain 10 lumen-seconds from this laser, it would have to be operated for a time of __________ seconds. — 100

(Note that a laser is a "powerful" source not because it produces a large flux, but because that flux is confined to a very small angle. Thus the intensity of the laser is high, but for all practical purposes in a single direction only.)

5. For pulsed sources (flashbulbs and electronic flash) the quantity of light they produce cannot be found by a simple multiplication as it can for a steady source like a tungsten lamp. The reason is that the flux from the lamp varies during the pulse, as shown by the curve repeated in the following diagram from the introduction to Section C3.

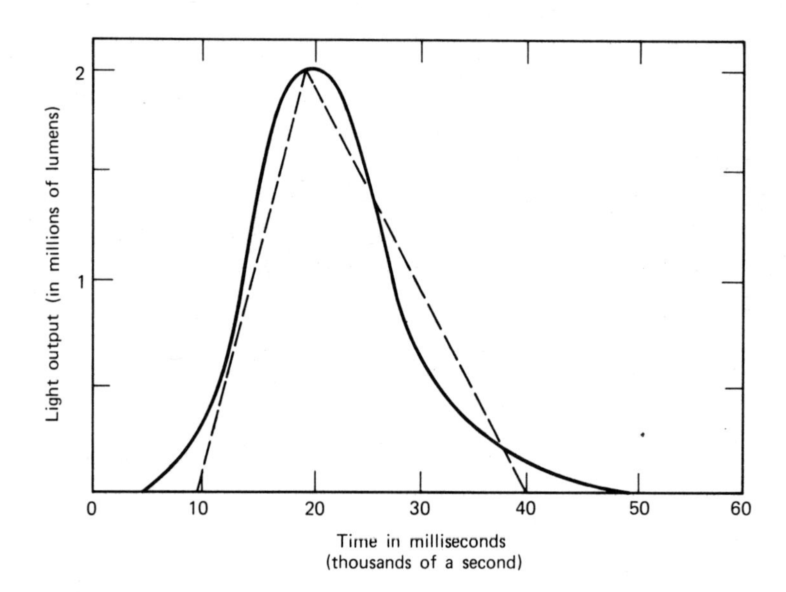

The flux, as shown on the vertical axis of the graph, begins at zero, rises to a very large value, and then falls again to zero as the lamp dies out. The *quantity* of light produced during the pulse is shown by the *area* under the curve. We can estimate this area (in lumen-seconds) if we note that the curve is not far from triangular, as indicated by the dotted lines on the graph. Recall that the area of a triangle is ½ of the product of the length of the base and the height (i.e., area = ½*bh*). Here the base is the time interval during which the lamp gives light, that is, from about 0.01 to 0.04 (10–40 *thousandths)* of a second, an interval of __________ second. — 0.03

6. The height of the peak of the curve is about __________ lumens. — 2 million

7. We now find the area by multiplying: ½ × 0.03 × 2,000,000 and get __________ lumen-seconds. — 30,000

8. We have found the total quantity of light produced by the lamp in all directions during the whole of the pulse. A different flashlamp has a peak output of 500,000 lumens. Its effective time interval during the pulse is 20 ms (20 thousandths of a second). Using the rule ½*bh,* the total quantity of light produced by the lamp is __________ lumen-seconds. — 5000

9. An electronic flashlamp produces light flux at a very high peak value—20,000,000 lumens. The time, however, is very short, perhaps only 1 thousandth of a second. The quantity of light in the pulse is __________ __________ .
number unit — 10,000 lumen-seconds

10. Even though the peak rate—the maximum flux—from the electronic flash is tremendous, the quantity of light is small because the time is so short. A 100-watt tungsten lamp produces light at a rate of about 2000 lumens. It could produce the

same quantity of light as the electronic flashlamp—10,000 lumen-seconds—but it would take __________ seconds to do so.

5

CAUTION. If only the number of lumen-seconds is reported for a pulsed lamp, one cannot know whether or not that lamp will be suitable for a specific photographic application.

REASONS.
1. The number indicates nothing about the direction in which the light is produced and therefore nothing about whether or not the light will fall on the subject properly.
2. The number of lumen-seconds gives equal weight to all of the light, even that in the tails of the curve where the light may be too feeble to produce an image on the film.

Even less meaningful is the somewhat similar term "beam candlepower seconds," often abbreviated BCPS. Because *beam candlepower* identifies only the *maximum intensity* of the lamp in some (unspecified) direction, it tells nothing about the distribution of light on the subject. You can have a very large value for BCPS if you are willing to have a very hot "hot spot" on the subject.

SELF-TEST ON MEASUREMENT OF LIGHT SOURCES—QUANTITY OF LIGHT (SECTION C4)

Check your understanding of this section by answering the following questions. The correct answers follow.

1. In what unit is the quantity of light from a source measured?
2. Which two factors determine the quantity of light from a source?
3. If the flux from a lamp is 500 lumens and it operates for a time of 5 seconds, what quantity of light is produced?
4. If 100 lumen-seconds is needed for some purpose, what is the time of operation needed if the lamp produces a flux of 50 lumens?
5. Estimate the total light output for the lamp represented by the uppermost curve in the set below:

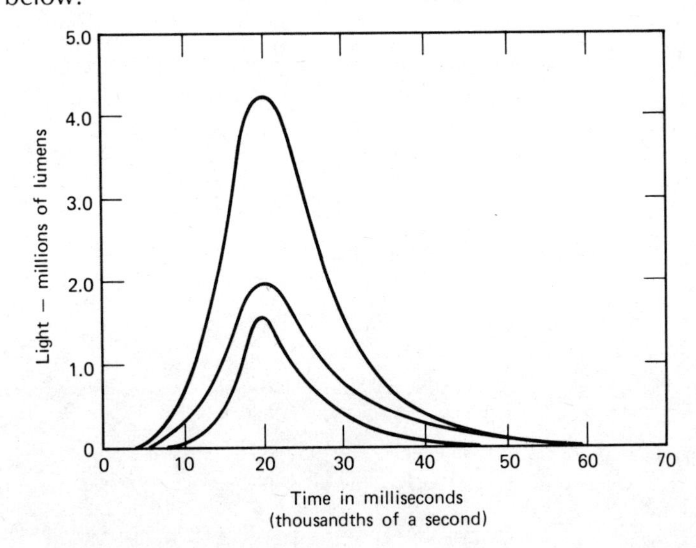

6. What are the two major deficiencies in the number found in (5)?

ANSWERS TO SELF-TEST ON MEASUREMENT OF LIGHT SOURCES—QUANTITY OF LIGHT (SECTION C4)

In the parentheses after each answer you find the items of this section that relate to the question and its answer.

1. The quantity of light from a source is measured in *lumen-seconds* (item 2).
2. The total quantity of light from a source is determined by the *flux* of the lamp in lumens and the *time* in seconds (item 1).
3. The quantity of light is the product of the flux (500 lumens) and the time (5 seconds); that is, 2500 lumen-seconds (item 3).
4. The time is the quotient of the quantity of light (100 lumen-seconds) divided by the flux (50 lumens); that is, 2 seconds (item 4).
5. The total light output is estimated from the area under the curve. Assume that it is nearly triangular; the base is from about 40 to 10 milliseconds, or 30 milliseconds, or 0.03 seconds. The altitude is about 4,100,000 lumens. The area of the triangle is ½ *bh*, or ½ × 0.03 × 4,100,000, or 61,500 lumen-seconds.
6. Not included in the total light output from a pulsed lamp are: (a) the variation in light with direction and (b) the flux variation during the pulse (text after item 10).

APPENDIX D

Measurement of Light Falling on a Surface

The intensity of a lamp remains the same in a given direction, no matter how far away it is. A candle is a candle even if it is a mile away.

What does usually change with distance is the light *falling on* an illuminated surface, such as the face of a model lighted with a studio lamp at different distances from the model. Guide numbers for flashlamps take this fact into account.

If you work carefully with the following items, you will learn:

1. The relationship between the intensity of a light source, its distance, and the light falling on an illuminated surface;
2. The important characteristic of a meter used to measure the light on a surface;
3. The relationship between the light falling on a film, the time, and the exposure.

DIRECTIONS. Cover about 2 inches of the right-hand margin of each of the following pages in turn with an opaque sheet of paper. Read carefully the first numbered stateement. *Write* the word or phrase that you believe correctly completes the statement. Move the cover sheet down to reveal the correct answer in the margin. Continue in this way until you have completed the section.

1. If a lamp of physically small area (a "point" source) has an intensity of 1 candela toward a surface, and if the source is 1 meter from the surface, the level of light on the surface is defined as 1 lux. (The older unit, equal to the lux, is called the "meter-candle.")

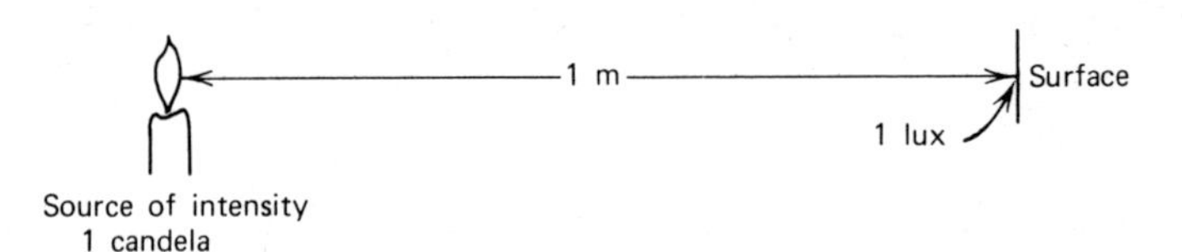

Note that the surface lies perpendicular to the light from the source. Note further that the size of the surface is irrelevant.

At a fixed distance the light level on the surface increases directly with the intensity of the source toward the surface, so with a *10*-candlepower lamp 1 meter from the surface, the light level on the surface would be ________ lux. — 10

2. The light level on the surface is called the *illuminance*. Illuminance is measured in ________ or in ________ . — lux, meter-candles (either order)
3. *Illuminance* needs to be distinguished from intensity. Recall that the intensity of a lamp is measured in ________ . — candles or candelas
4. Illuminance must be distinguished from luminance. Recall that *luminance* applies to a *source* of light, and is measured in candelas per unit ________ of the source. — area
5. *Illuminance* is a measure of the light *on* a ________ . — surface
6. *Intensity* and *luminance* are measures of the light *from* a ________ . — source, lamp (etc.)
7. If a lamp has an *intensity* of 50 candelas, and if it is 1 meter from a surface, the surface receives an illuminance of ________ lux (meter-candles). — 50
8. As the distance between a *point* source of light and an illuminated surface changes, the illuminance on the surface also changes. As the distance to the surface *increases*, the illuminance on the surface ________ . — decreases or is less

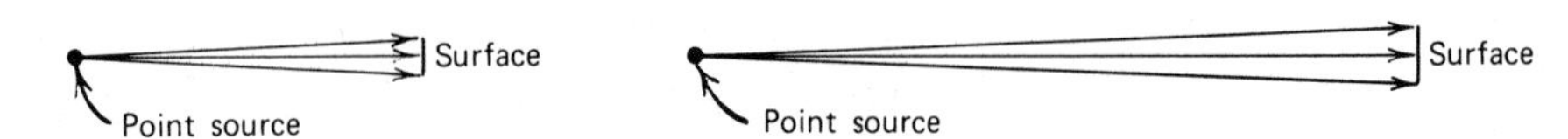

9. The sketch that follows, repeated from Section C3, item 2, shows that the light on a

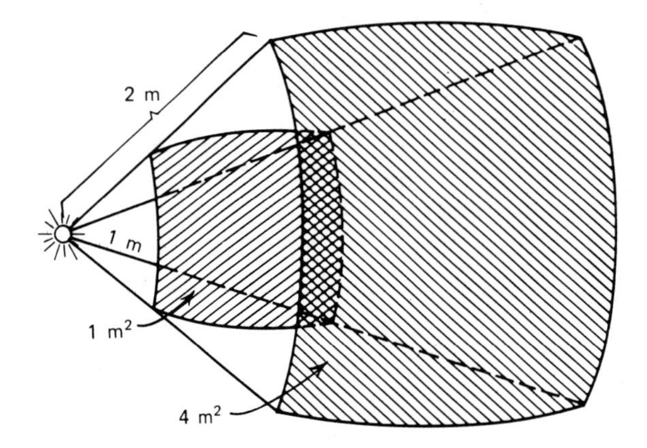

surface 1 meter square is spread out over 4 m² if the distance is increased from 1 to ________ meters. — 2

10. Since at the greater distance the light is distributed over *four* times as much area, the light on *each* square meter would, at the greater distance, be only one-________ th of what it was at the lesser distance. — (one) four(th)

11. If the surface were moved to 3 meters away, the covered area would be ________ square meters. — 9
12. Now the light on each square meter would be only one-________ of what it was at a distance of 1 meter. — ninth
13. The light level on each square meter is in fact the *illuminance* on the surface. If the distance increases, the illuminance ________. — decreases
14. The change in illuminance is related to the *square* of the distance. As the distance is doubled (increased by a factor of 2), the illuminance is reduced by a *divisor* of ________. — 4
15. If the distance were increased by a factor of 10, the illuminance would decrease by a divisor of ________. — 100
16. Conversely, as the distance is reduced, the illuminance is ________. — increased or greater (etc.)
17. If the distance were cut in half (reduced by a divisor of 2), the illuminance would increase by a factor of ________. — 4
18. The preceding examples illustrate the "inverse square" law of illumination. The law applies strictly only to a small-area source, in other words, a ________ source. — point
19. The word "inverse" means that the illuminance changes oppositely in direction to the change of the ________. — distance
20. Thus if we compare the English unit foot-candle with the meter-candle, the sketch below indicates the situation:

   1 meter   1 foot

   Since for 1 foot-candle the source is *closer* to the surface than for 1 meter-candle, 1 foot-candle represents a ________ illuminance than does 1 meter-candle. — greater, or larger
21. The word "square" in the name of the inverse-*square* law means that the values of the distances involved must be multiplied by the same values to ascertain how the illuminance changes. It is easier to find the ratio of the distances and then square that ratio. To find the relationship between a foot-candle and a meter-candle, we use the fact that 1 meter is about 3.28 feet. The square of 3.28 is about 10.8. Thus one foot-candle is approximately ________ meter-candles (lux). — 10.8
22. Thus to change a value in foot-candles to the equivalent value in lux, multiply by ________. — 10.8
23. A surface is illuminated by a point source. The distance is changed from 3 to 12 meters. To find the change in the illuminance on the surface, we find first the ratio of the distances. It is ________ to 1. — 4
24. We then square this ratio and get ________. — 16
25. The illuminance changes inversely. Thus we know that at 12 meters the illuminance will be one-________ of that at 3 meters. — sixteenth
26. The inverse-square law is expressed by this formula: $E = I/d^2$, where E is the illuminance on the surface, I is the intensity of the point source toward the surface, and d is the distance of the point source from the surface. E is in lux (meter-

candles) if *I* is in candelas and *d* is in meters. *E* is in foot-candles if *I* is in candelas and *d* is in feet.

From the formula, if a point source has an intensity of 500 candelas and is 10 meters from a surface, the illuminance on the surface is ________ ___ .
number unit

5 lux, or 5 meter-candles

27. If a point source of intensity 400 candelas is 2 feet from a surface, the illuminance on the surface is ________ ___ .
number unit

100 foot-candles

28. From the same formula, if an illuminance of 2 lux is needed on a surface 4 meters from a point source, the intensity of the source must be ________ ___ .
number unit

32 candelas

This rule applies *only* if the source is small (its largest dimension should be no greater than one-tenth of the distance to the surface) and the surface must be perpendicular to the direction of the light from the source.

29. Meters that measure illuminance are commonly called "incident-light" meters. As distinguished from *luminance* meters (Section C2, item 21), illuminance meters must record light from whatever direction it comes to the surface and thus must have an angle of view of 180°. Thus of the two meters sketched below, meter ____________ has an angle of view close to 180°.

B

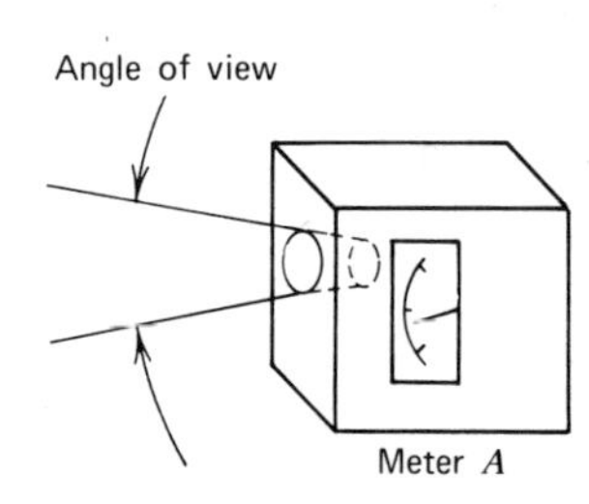

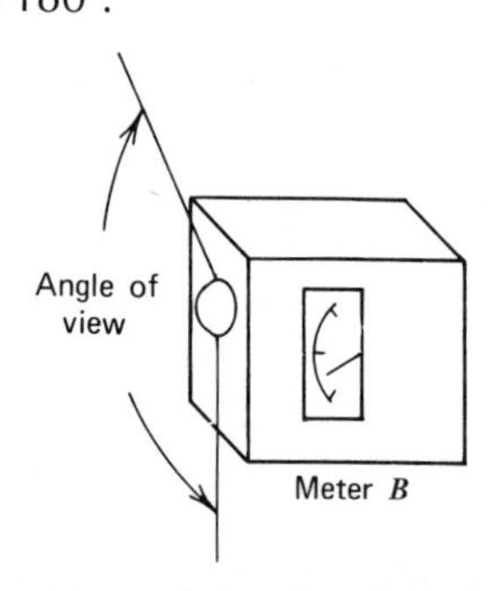

30. Therefore meter *B* is an ____________ -meter. Meter *A* is a luminance-meter.

illuminance

31. When the photographer uses a *luminance*-meter, he aims the meter at the ____________ (Section C2, item 21).

subject

32. On the other hand, an illuminance-meter is placed at the *surface* of interest, with the photosensitive cell lying in the same attitude as the surface, so that it receives light as does the surface.

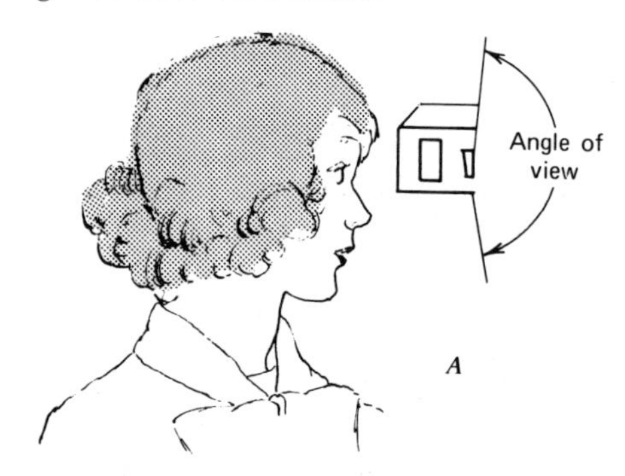

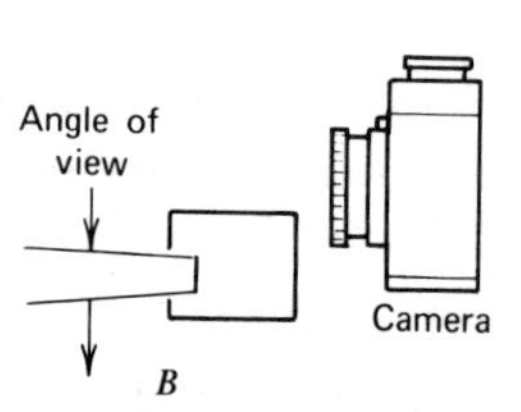

Thus to measure the illuminance *on* the forehead of the model in the sketch above—an incident-light reading—the meter is held as shown at letter ____________ .

A

33. For such a reading to be reliable, the angle of view of the meter needs to be close to ________ degrees. — 180

34. On the other hand, for a luminance—a "reflected-light" reading—the meter needs to be placed as shown at letter ________ . — *B*

35. For a luminance reading, the angle of view of the meter needs to be ________ . (small, large) — small

36. Cameras equipped with inside-the-camera light meters are arranged so that the meter faces the lens, and thus estimates the light falling on the exposure plane. Such a meter is estimating ________ . — illuminance

37. In monitoring a projection printer for uniformity of light *on* the easel, a(n) ________ -meter is needed. — illuminance

38. On the other hand, in checking the uniformity of a motion-picture screen as seen by the audience, the apparent brightness of the screen is what is important. For this purpose, one would use a(n) ________ -meter. — luminance

To complete this treatment of the basic concepts involved in light measurement, the following items relate to the meaning of exposure. See also Chapter 2, items 51–83, for a fuller treatment of exposure.

39. Illumination on a surface, measured by the illuminance in lux is (like flux from a source) a *rate* process. Think of illuminance as measuring the rate of reception of light (like a shower of water) on a surface. The *amount* (quantity) of light received by a given area of photosensitive material will be determined both by the illuminance (the rate) and the ________ during which light falls on the surface. — time

40. You may cause a film to receive a large quantity of light by supplying the film with light at a high rate (great illuminance) over a short time or with a weak illuminance over a ________ time. — long

41. The quantity of light is found from the product of the illuminance and the time: $H = Et$, where H stands for the amount of exposure, E for the illuminance, and t for the time. H comes out in lux-seconds (or meter-candle-seconds, which mean the same) if the illuminance E is measured in lux (meter-candles) and the time in seconds. If the illuminance on the film is 5 lux and the time is 4 seconds, the exposure is ________ lux-seconds. — 20

42. From the defining formula $H = Et$, if an exposure of 100 lux-seconds is needed and the illuminance is 50 lux, the required time is ________ seconds. — 2

43. The same exposure (100 lux-seconds) could be produced by many different combinations of illuminance and time. If the time had to be 1 second, the illuminance would have to be ________ ____ . (number unit) — 100 lux

44. To get an exposure of 100 lux-seconds with a time of 0.1 second, the illuminance would have to be ________ ____ . (number unit) — 1000 lux

45. If you could afford a very long exposure time, such as 1000 seconds, an exposure of 100 lux-seconds could be obtained with an illuminance of only ________ ____ . (number unit) — 0.1 lux

46. In the camera, the image of the subject (as produced by the camera lens) consists of different *illuminances*. Each different illuminance is associated with a different

subject *luminance* (see Section C2). The time of exposure is usually the same all over the film, but because the illuminances are different, there will be different __________ at different places on the film. exposures

For any ordinary subject, there will be a very large number of different illuminances on the film, and thus a very large number of different exposures.

SELF-TEST ON MEASUREMENT OF LIGHT FALLING ON A SURFACE

Check your understanding of this section by answering the following questions. The correct answers follow.

1. The technical term for the level of light on a surface is __________ .
2. The unit of measure of the strength of light on a surface is __________ or __________ - __________ .
3. The unit for the intensity of a lamp is __________ .
4. The luminance of a source is measured in candelas per unit __________ .
5. If a photographer needs to measure the light *on* a copy board, he needs to use a(n) __________ meter.
6. According to the inverse-square law, doubling the distance of a surface from a small lamp causes the illuminance on the surface to become __________ .
7. Doubling the distance in this case would cause the illuminance to become __________ - __________ as much as before.
8. If instead the distance were made one quarter as great as in the first case, the illuminance would become __________ .
9. Reducing the distance to one quarter of its original value would cause the illuminance to become __________ times as great.
10. By the inverse-square formula $E = I/d^2$, if a small lamp of intensity 100 candelas is 2 meters from a surface, the illuminance on the surface is ______ ____ .
    number unit
11. Some light meters have the photocell on a cord separated from the dial. Which of the following is an illuminance-meter?

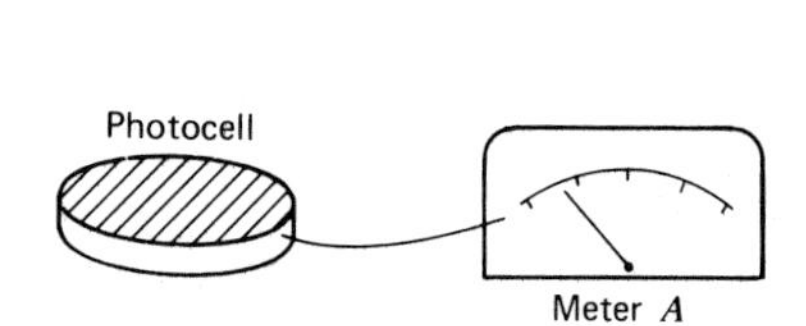

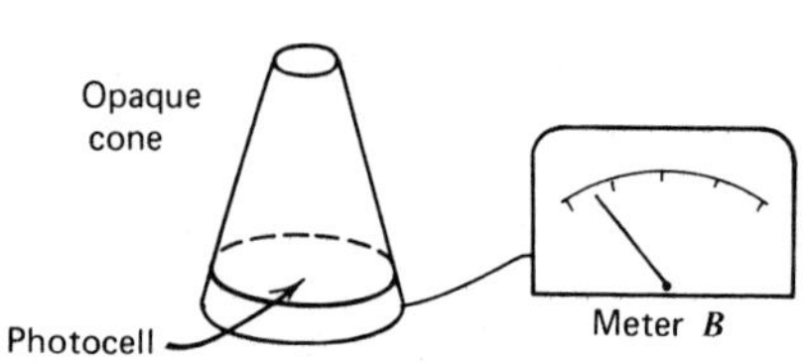

How should the meter be aimed (as indicated by the arrow in each case) to measure the illuminance on the door of the house in the sketch on page 217?

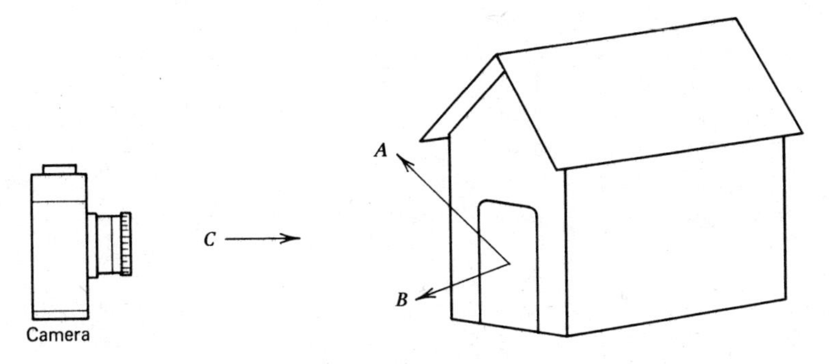

13. From the formula $H = Et$, if the illuminance on the easel of a projection printer is 250 lux and the exposure time is 5 seconds, the exposure is _______ ___ .
number unit

14. The subject of a photograph is a landscape. Using camera settings of 1/50 at $f/16$, there would be on the film in the camera ______________ different exposure(s).
one, a few, many

ANSWERS TO SELF-TEST ON MEASUREMENT OF LIGHT FALLING ON A SURFACE

In the parentheses after each answer you find the numbers of the items in this section that relate to the question and its answer.

1. Illuminance (item 2).
2. Lux or meter-candles or foot-candles (items 1, 2, 20, 22).
3. Candles or candelas (item 3).
4. Area (item 4).
5. Illuminance (item 5).
6. Less, or smaller (items 9–10).
7. One-fourth (items 10, 14).
8. Larger, or greater (items 16, 17).
9. 16 (items 16, 17).
10. 25 lux (or meter-candles) (items 26–28).
11. Meter *A* (items 29–30).
12. Situation *B* (items 31–32).
13. 1250 lux-seconds (or meter-candle-seconds) (items 41–43).
14. Many (item 46).

Annotated Bibliography

Fundamentals of Photography. Neblette, Van Nostrand-Rheinhold, N.Y. An elementary text containing a brief review of basic sensitometry.

Basic Photo Series. Adams, Morgan and Morgan, Dobbs Ferry, N.Y. The application of sensitometric principles to creative photography.

The New Zone System Manual. White, Zakia and Lorenz, Morgan and Morgan, Dobbs Ferry, N.Y. The application of sensitometric methods and principles to in-camera tests of a photographic system, and the application of the results for making more effective photographs.

Zone Systemizer. Dowdell and Zakia, Morgan and Morgan, Dobbs Ferry, N.Y. A device for aiding the photographer in previsualizing the photograph based on sensitometric principles.

101 Experiments in Photography. Zakia and Todd, Morgan and Morgan, Dobbs Ferry, N.Y. Simple, but useful methods for testing and evaluating photographic materials, as well as suggestions for other experiments.

Photographic Sensitometry. Todd and Zakia, Morgan and Morgan, Dobbs Ferry, N.Y. Chapters 1–5 and 8 cover in text form the material of this programmed book. The remaining chapters treat special topics in sensitometry.

Theory of the Photographic Process. James (Ed.), Macmillan, N.Y. Written primarily for scientists, an authoritative work, including chapters on sensitometry and tone reproduction.

Handbook for Photographic Engineers. Woodlief (Ed.), Society of Photographic Scientists and Engineers, Washington, D.C. A reference book containing massive amounts of data relating to all areas of engineering photography.

Photographic manufacturers —du Pont, Eastman Kodak, General Aniline and Film (etc.)—publish sensitometric data for their products. Camera stores can supply additional information.

The American Standards Institute. N.Y., can supply test methods in sensitometric, as well as other, photographic fields.

Index

Numbers in parentheses following the page numbers refer to the items for the programmed sections.